The Sciences Po Series in International Relations and Political Economy

Series Editor, Christian Lequesne

This series consists of works emanating from the foremost French researchers from Sciences Po, Paris. Sciences Po was founded in 1872 and is today one of the most prestigious universities for teaching and research in social sciences in France, recognized worldwide.

This series focuses on the transformations of the international arena, in a world where the state, though its sovereignty is questioned, reinvents itself. The series explores the effects on international relations and the world economy of regionalization, globalization (not only of trade and finance but also of culture), and transnational flows at large. This evolution in world affairs sustains a variety of networks from the ideological to the criminal or terrorist. Besides the geopolitical transformations of the globalized planet, the new political economy of the world has a decided impact on its destiny as well, and this series hopes to uncover what that is.

Published by Palgrave Macmillan:

Politics In China: Moving Frontiers
  edited by Françoise Mengin and Jean-Louis Rocca
Tropical Forests, International Jungle: The Underside of Global Ecopolitics
  by Marie-Claude Smouts, translated by Cynthia Schoch
The Political Economy of Emerging Markets: Actors, Institutions and Financial Crises in Latin America
  by Javier Santiso
Cyber China: Reshaping National Identities in the Age of Information
  edited by Françoise Mengin
With Us or Against Us: Studies in Global Anti-Americanism
  edited by Denis Lacorne and Tony Judt
Vietnam's New Order: International Perspectives on the State and Reform in Vietnam
  edited by Stéphanie Balme and Mark Sidel
Equality and Transparency: A Strategic Perspective on Affirmative Action in American Law
  by Daniel Sabbagh, translated by Cynthia Schoch and John Atherton
Moralizing International Relations: Called to Account
  by Ariel Colonomos, translated by Chris Turner
Norms over Force: The Enigma of European Power
  by Zaki Laidi, translated by Cynthia Schoch
Democracies at War against Terrorism: A Comparative Perspective
  edited by Samy Cohen, translated by John Atherton, Roger Leverdier, Leslie Piquemal, and Cynthia Schoch

Justifying War? From Humanitarian Intervention to Counterterrorism
  edited by Gilles Andréani and Pierre Hassner, translated by John Hulsey, Leslie
  Piquemal, Ros Schwartz, and Chris Turner
An Identity for Europe: The Relevance of Multiculturalism in EU Construction
  edited by Riva Kastoryano, translated by Susan Emanuel
*The Politics of Regional Integration in Latin America: Theoretical and Comparative
Explorations*
  by Olivier Dabène
Central and Eastern Europe: Europeanization and Social Change
  by François Bafoil, translated by Chris Turner
Building Constitutionalism in China
  edited by Stéphanie Balme and Michael W. Dowdle
In the Name of the Nation: Nationalism and Politics in Contemporary Russia
  by Marlène Laruelle
Organized Crime and States: The Hidden Face of Politics
  edited by Jean-Louis Briquet and Gilles Favarel-Garrigues
Israel's Asymmetric Wars
  by Samy Cohen, translated by Cynthia Schoch
China and India in Central Asia: A New "Great Game"?
  edited by Marlène Laruelle, Jean-François Huchet, Sébastien Peyrouse, and
  Bayram Balci

China and India in Central Asia

A New "Great Game"?

Edited by
Marlène Laruelle, Jean-François Huchet,
Sébastien Peyrouse, and Bayram Balci

CHINA AND INDIA IN CENTRAL ASIA
Copyright © Marlène Laruelle, Jean-François Huchet, Sébastien Peyrouse, and Bayram Balci, 2010.

First published in 2010 by
PALGRAVE MACMILLAN®
in the United States—a division of St. Martin's Press LLC,
175 Fifth Avenue, New York, NY 10010.

Where this book is distributed in the UK, Europe and the rest of the world, this is by Palgrave Macmillan, a division of Macmillan Publishers Limited, registered in England, company number 785998, of Houndmills, Basingstoke, Hampshire RG21 6XS.

Palgrave Macmillan is the global academic imprint of the above companies and has companies and representatives throughout the world.

Palgrave® and Macmillan® are registered trademarks in the United States, the United Kingdom, Europe and other countries.

ISBN: 978–0–230–10356–6

Library of Congress Cataloging-in-Publication Data

China and India in Central Asia : a new "great game"? / edited by Marlène Laruelle ... [et al.].
    p. cm.—(Sciences PO series in international relations and political economy)
  Includes bibliographical references.
  ISBN 978–0–230–10356–6 (alk. paper)
   1. China—Foreign relations—Asia, Central. 2. Asia, Central—Foreign relations—China. 3. India—Foreign relations—Asia, Central. 4. Central, Asia—Foreign relations—India. 5. Geopolitics—Asia, Central. I. Laruelle, Marlène.

JZ1734.A55C45 2010
958'.043—dc22   2010013331

A catalogue record of the book is available from the British Library.

Design by Newgen Imaging Systems (P) Ltd., Chennai, India.

First edition: November 2010

10 9 8 7 6 5 4 3 2 1

Printed in the United States of America.

CONTENTS

List of Illustrations   vii

Notes on Contributors   ix

1   Why Central Asia? The Strategic Rationale of Indian and Chinese
    Involvement in the Region   1
    The Editors

Part I   Negotiating Projections of Power in Central Asia

2   Russia Facing China and India in Central Asia: Cooperation,
    Competition, and Hesitations   9
    Marlène Laruelle

3   Central Asia–China Relations and Their Relative
    Weight in Chinese Foreign Policy   25
    Jean-Pierre Cabestan

4   An Elephant in a China Shop? India's Look North to Central
    Asia...Seeing Only China   41
    Emilian Kavalski

5   Afghanistan and Regional Strategy: The India Factor   61
    Meena Singh Roy

6   Afghan Factor in Reviving the Sino-Pak Axis   81
    Swaran Singh

Part II   India and China in Central Asia, between Cooperation, Parallelism, and Competition

7   India and China in Central Asia: Mirroring Their
    Bilateral Relations   97
    Jean-François Huchet

8 India–China Interactions in Central Asia through
   the Prism of Paul Kennedy's Analysis of Great Powers   117
   Basudeb Chaudhuri and Manpreet Sethi

9 Cooperation or Competition? China and India in Central Asia   131
   Zhao Huasheng

Part III   Chinese and Indian Economic Implementations from the Caspian Basin to Afghanistan

10 Scramble for Caspian Energy: Can Big Power Competition
    Sidestep China and India?   141
    P. L. Dash

11 Comparing the Economic Involvement of China and India
    in Post-Soviet Central Asia   155
    Sébastien Peyrouse

12 The Reconstruction in Afghanistan: The Indian and Chinese
    Contribution   173
    Gulshan Sachdeva

Part IV   Revisited Historical Backgrounds, Disputed Religious Modernities

13 From the Oxus to the Indus: Looking Back at India–Central
    Asia Connections in the Early Modern Age   197
    Laurent Gayer

14 Uyghur Islam: Caught between Foreign Influences and
    Domestic Constraints   215
    Rémi Castets

15 The Jama'at al Tabligh in Central Asia—a Mediator in the
    Recreation of Islamic Relations with the Indian Subcontinent 235
    Bayram Balci

Index   249

ILLUSTRATIONS

Tables

5.1	India–Afghanistan Trade	71
8.1	China–India Global Comparison	119
11.1	Chinese and India Bilateral Trade with Central Asia in 2008	156
12.1	Some Macroeconomic Indicators in Afghanistan	176
12.2	Coalition Military Fatalities in Afghanistan, 2001–2009	177
12.3	U.S. Government Funding Provided in Support of Afghan Security, Stabilization, and Development, Fiscal Years 2002–2009	179
12.4	Afghan Population's Opinion about Different Countries, 2009	183
12.5	Afghan Population's Opinion about Overall Role Played by Different Countries, 2009	183

Graph

7.1	Evolution of India–China Bilateral Trade	107

CONTRIBUTORS

Bayram Balci is a Director of the French Institute for Central Asian Studies since 2006. Between 2001 and 2006 he was a researcher and coordinator of the Caucasus Program based in Baku, Azerbaijan, for the French Institute on Anatolian Studies. His personal field and research work focus on religious globalization, Islamic sociology and education, and migrations and pilgrimages in the Turkic world. He has published *Missionnaires de l'Islam en Asie centrale, les écoles turques de Fethullah Gülen* (Islamic Missionaries in Central Asia: The Turkish Schools of Fethullah Gülen), and edited *Religion, Société et politique dans le Caucase post-soviétique* (Religion, Society and Politics in the Post-Soviet Caucasus).

Rémi Castets defended his PhD on the Uyghur question in 2010 at Sciences Po. Between 2004 and 2006, he worked as a research fellow at the French Centre for Research on Contemporary China (CEFC, Hong Kong). Since 2006, he has been Lecturer at the University Michel de Montaigne Bordeaux 3 where he teaches geopolitics and Chinese political history. He is also junior researcher associated to the CERI (Centre for International Studies and Research, Paris).

Jean-Pierre Cabestan is Professor and Head of the Department of Government and International Studies at Hong Kong Baptist University. He is also associate researcher at the Asia Centre at Sciences Po, Paris. His most recent publications include (with Benoît Vermander) *La Chine en quête de ses frontières. La confrontation Chine-Taiwan* (Paris: Presses des Sciences Po, 2005), translated and published in Chinese as a special issue of the journal *Renlai* (Taipei) in January 2007, *La politique internationale de la Chine. Entre intégration et volonté de puissance* (Paris: Presses de Sciences-Po, 2010), and, as a co-author, *La Chine et la Russie: entre convergences et méfiance* (Paris: Unicomm, 2008). He has also published numerous articles in English on China's political system and reform, Chinese law, the relations across the Taiwan Strait, and Taiwanese politics. He received his PhD from the University of Paris 1, Panthéon-Sorbonne.

Basudeb Chaudhuri is an economist trained at the Presidency College (Kolkata), the Indian Statistical Institute, and the University of Paris I Panthéon Sorbonne. An Associate Professor of Economics and a former Vice President of the University of Caen, Normandy, he is currently on lien in Delhi as Director of the Centre de Sciences Humaines, a research unit of the French External affairs Ministry and CNRS. He has edited, with Frédéric Landy, *Globalization and Local Development in India: Examining the Spatial Dimension* (New Delhi: Manohar Publishers—Centre de Sciences Humaines, 2004). He has written chapters and articles on the Indian economy in the *Oxford Companion to Economics in India* (edited by Kaushik Basu) (Oxford, UK: Oxford University Press, 2007), the *Dictionnaire de l'Inde* (edited by C. Clementin Ojha, C. Jaffrelot, D. Matringe, and J. Pouchepadass) (Paris: Larousse, 2009), and in other international journals and books. His areas of interest include developing and emerging economies, public economics, institutional economics and political economy, and globalization.

P. L. Dash is Professor of Eurasian Studies at the Centre for Central Eurasian Studies, University of Mumbai, India. His areas of interest are post-Soviet development in Russia, sociopolitical changes in Central Asia, Caspian oil politics, and Indo-Russian Relations. Two of his recent publications are *Indo-Russian Relations: Sixty Years of Enduring Legacy* (co-edited with Andrei M. Nazarkin) (New Delhi: Academic Excellence Publisher, 2008) and *Caspian Pipeline Politics, Energy reserves and Regional Implications* (New Delhi: Pentagon Press in association with Observer Research Foundation, 2008). His publications include 11 books and over 120 research articles.

Laurent Gayer is a research fellow at CNRS, attached to the Centre universitaire de recherches sur l'action publique et le politique (CURAPP), Amiens, presently posted at the Centre de sciences humaines (CSH), New Delhi. After completing a PhD in international relations at Sciences Po, Paris, he has been focusing on the political sociology of irregular armed forces in India and Pakistan. His recent publications include *Armed Militias of South Asia. Fundamentalists, Maoists and Separatists* (co-edited with Christophe Jaffrelot) (London/New York: Hurst/Columbia University Press, 2009).

Emilian Kavalski (PhD, Loughborough University) is Lecturer in Politics and International Relations at the University of Western Sydney (Australia). He has held Marie Curie research positions at Aalborg University (Denmark) and Ruhr University-Bochum (Germany), the I. W. Killam Postdoctoral Fellowship at the University of Alberta (Canada), and the Andrew Mellon Fellowship at the American Center for Indian Studies (New Delhi, India). Dr. Kavalski's current research deals with the complexity of security governance and interactions between China, India, and the European Union in

Central Asia. He recently authored *India and Central Asia: The Mythmaking and International Relations of a Rising Power* (London, UK: I.B.Tauris, 2010) and edited *Stable Outside, Fragile Inside? Post-Soviet Statehood in Central Asia* (Aldershot, UK: Ashgate, 2010), *The "New" Central Asia: The Regional Impact of International Actors* (Hackensack, NJ: World Scientific, 2010), *China and the Global Politics of Regionalization* (Aldershot, UK: Ashgate, 2009).

Jean-François Huchet is currently director of the French Centre for Research on Contemporary China (CEFC) in Hong Kong, editor of the review *China Perspectives* (and its French edition *Perspectives Chinoises*), and Associate Professor of Economics at the University of Rennes 2 in France. He was based in Asia for 15 years and has occupied several academic positions at the Maison franco-japonaise in Tokyo, CEFC in Hong Kong, and Peking University. He has received his PhD in economics from the University of Rennes 1. He has published numerous articles and books on the Chinese economy, especially on the reform of the state-owned enterprises. He has recently edited, with Wang Wei, *Chinese Firms in the Era of Globalisation* (in Chinese and English) (Beijing: Zhongguo Fazhan Chubanshe, 2008) and with Joël Ruet and Xavier Richet, *Globalisation of Firms in China, India and Russia* (New Delhi: Academic Foundation, 2007).

Marlène Laruelle is a Senior Research Fellow with the Central Asia-Caucasus Institute and Silk Road Studies Program, a joint center affiliated with Johns Hopkins University's School of Advanced International Studies, Washington DC, and the Institute for Security and Development Policy, Stockholm. In Paris, she is an Associate Scholar at the French Center for Russian, Caucasian and East European Studies at the School of Advanced Social Sciences Studies (EHESS), and at the Post-Soviet Studies Department at Sciences Po. Her main areas of expertise are nationalism, national identities, political philosophy, intellectual trends, and geopolitical conceptions of local elites in Russia and Central Asia. She has expertise in Russian and Central Asian foreign policy, and in Russian policy toward Central Asia. Her English-language publications include *Russian Eurasianism. An Ideology of Empire* (Washington DC: Woodrow Wilson Press/Johns Hopkins University Press, 2008); *In the Name of the Nation. Nationalism and Politics in Contemporary Russia* (New York: Palgrave Macmillan, 2009); and, as editor, *Russian Nationalism and the National Reassertion of Russia* (London: Routledge, 2009).

Sébastien Peyrouse is a Senior Research Fellow with the Central Asia-Caucasus Institute and Silk Road Studies Program, a joint center affiliated with Johns Hopkins University's School of Advanced International Studies, Washington DC, and the Institute for Security and Development Policy, Stockholm. He was a doctoral and postdoctoral Fellow at the French

Institute for Central Asia Studies in Tashkent (1998–2000 and 2002–2005), a Research Fellow at the Slavic Research Center, Hokkaido University in Sapporo (2006), and a Research Fellow at the Woodrow Wilson International Center for Scholars in Washington (2006–2007). In Paris, he is an Associated Fellow at the Institute for International and Strategic Relations. His research originally focused on the impact of the Russian/Soviet heritage in the five Central Asian republics. His main areas of expertise are political systems in Central Asia, Islam and religious minorities, and Central Asia's geopolitical positioning toward China, Russia, and South Asia. Peyrouse is the author or co-author of six French books on Central Asia. In English, he has published *China as a Neighbor. Central Asian Perspectives and Strategies* (Washington, DC: Central Asia-Caucasus Institute, 2009) with Marlène Laruelle, and "The Economic Aspects of the Chinese-Central-Asia Rapprochement" (*Silk Road Papers*, Central Asia-Caucasus Institute, 2007).

Gulshan Sachdeva is Associate Professor at the School of International Studies, Jawaharlal Nehru University, New Delhi. As a regional cooperation adviser, he has worked with the Asia Foundation and with the Asian Development Bank in Kabul and implemented projects at the Ministry of Foreign Affairs, Afghanistan. He was Visiting Professor at the University of Antwerp, University of Trento, and Corvinus University of Budapest and also Visiting Fellow at the Institute of Oriental Studies Moscow, Institute of Oriental Studies Almaty, and at the Cambridge Central Asia Forum. He is author of *Economy of the Northeast* (New Delhi: Konark Publishers, 2000), various monographs, project reports, and 55 research papers in scholarly journals and edited books. He is also member of the governing board for the India-Central Asia Foundation. He holds a PhD in Economic Science from the Hungarian Academy of Sciences.

Manpreet Sethi heads the project on Nuclear Security at the Centre for Air Power Studies (CAPS), New Delhi. She is also Fellow, International Relations, Centre de Sciences Humaines, New Delhi. Over the last twelve years, since completion of her PhD from the Latin American Division of the School of International Studies, Jawaharlal Nehru University, she has focused on issues related to nuclear strategy, energy, proliferation, export controls, and disarmament. She was earlier on the research faculty of the Institute for Defence Studies and Analyses, New Delhi, from 1997–2001. She is author of *Nuclear Strategy: India's March towards Credible Deterrence* (New Delhi: Knowledge World, 2009) and *Argentina's Nuclear Policy* (New Delhi: Knowledge World, 1999), co-author of *Nuclear Deterrence and Diplomacy* (New Delhi: Knowledge World, 2004), and editor of *Global Nuclear Challenges* (New Delhi: Knowledge World, 2009) and *Towards a World Free of Nuclear Weapons* (New Delhi: Knowledge World, 2009). She is also author of an Occasional

Paper entitled *"Nuclear Deterrence in Second Tier States: A Case Study of India"* (New Delhi: CSH, 2009). Her research papers are widely published in national and international academic journals and books.

Swaran Singh teaches Disarmament Studies at the School of International Studies, Jawaharlal Nehru University (New Delhi). He is President of the Association of Asian Scholars (an Asia-wide network in New Delhi), General Secretary of the Indian Association of Asian and Pacific Studies (headquartered in Varanasi), and a member of the Bangkok-based Asian Scholarship Foundation's Regional Review Committee for South Asia. He has traveled and written extensively on Asian affairs and China's foreign and security policy issues with a special focus on China–India confidence-building measures as also on India's foreign and security policy issues. More recently, he has authored *China-India Economic Engagement: Building Mutual Confidence* (2005), *China-South Asia: Issues, Equations, Policies* (2003), *China's Changing National Security Doctrines* (1999) and *Limited War* (1995); and recently edited the compilation *China-Pakistan Strategic Cooperation: Indian Perspectives* (2007) and co-authored *Regionalism in South Asian Diplomacy* (SIPRI Policy Paper No. 15, February 2007). He is currently working on a monograph titled *Nuclear Command and Control in Southern Asia: China, India, Pakistan*.

Meena Singh Roy is a Research Fellow at the Institute for Defence Studies and Analyses. Her area of specialization is Central Asia, Russia, Iran, and Southern Africa. She completed her PhD from the University of Delhi in 1994. She has been a senior research scholar in the Department of African Studies, Delhi University. She has also been associated with the Institute of Commonwealth Studies, School of Oriental and African Studies, and the London School of Economics for her research work. She has presented papers in various national and international seminars. She has published various research papers and articles in referred journals and books. She has been involved in publishing IDSA energy newsletter. Her last publication was *International and Regional Security Dynamics: Indian and Iranian Perspectives* (ed.) (New Delhi: Institute for Defence Studies and Analyses, July 2009). Currently she is working on a book titled *Reshaping India—Central Asia Relations in the New Strategic Environment*.

Zhao Huasheng is Director of the Center for Russian and Central Asian Studies, Fudan University, Shanghai, Popular Republic of China.

CHAPTER 1

Why Central Asia? The Strategic Rationale of Indian and Chinese Involvement in the Region

THE EDITORS

Since the fall of the Soviet Union, the rediscovery of Central Asia by the international community has placed this region in a specific intellectual context, one marked by a return of geopolitical theories and debates around the "end of history" and the "clash of civilizations." The revival of geopolitical theory, especially Sir Halford Mackinder's idea that one who controls the *Heartland* controls the world, has profoundly shaped the new frameworks applied to the post-Soviet states of Central Asia and to Afghanistan. In contrast to the geographical and economic isolation of the region, theories about the revival of the Silk Road flourished in the West and in Asia. The United States and the European Union have used them to promote the release of Central Asia from the Russian sphere of influence by opening toward the south. Turkey, Iran, Japan, South Korea, China, India, and Pakistan have made references to their historical ties with the region, beyond the years of the Iron Curtain.

Although the fall of the Soviet Union took the entire international community by surprise, it has drastically changed the geopolitical situation in China and India. The former saw the collapse of its main enemy from the 1960s and 1970s and discovered a new area of potential instability on its north and northwestern borders. The Chinese authorities, unprepared and worried about the possible repercussions of this historic event on their political system and territorial unity, implemented an active "good neighborhood" policy with Russia and Central Asia. Less than two decades later, Moscow and Beijing have signed a strategic partnership, as have Astana and Beijing. China has become an indispensable diplomatic and economic ally of the post-Soviet states, multilateral cooperation mechanisms have been developed, new cultural interactions have emerged, and popular concerns have taken shape in Russia and in Central Asia about the future of the Chinese presence. For India too, the situation has changed, but in a different way. The loss of the Soviet ally has undermined the political and economic choices of the Indian regime since the departure of Britain, forcing a complex international reorientation marked by a fear of the growing Sino-Pakistani alliance and the development of a new dialogue with the United States.

Direct Indian-Central Asian links were limited during the Soviet period, but the context of Indian-Soviet friendship made Delhi relatively present in the everyday lives of Central Asians via television, movies, music, and cultural exchanges. China has inevitably, albeit cumbersomely, passed from the status of historic enemy to that of partner. Meanwhile India has lost relative visibility since the disappearance of the Soviet Union and is now trying to gain in the strategic sector what it has lost in its cultural presence. In post-Soviet Central Asia as in Afghanistan, people have a positive vision of the Indian presence, whether through historical memory, a sense of cultural proximity, or political sympathy. The relation to China is much more complex, dotted with Sinophobic clichés linked to the myth of the "yellow peril" or denunciations of the implementation methods of Chinese companies. This dissociation is nothing specific to Central Asia. In the West too, India elicits less concern than China, not only because of a view based on cultural and political arguments, but also and especially because an Indian superpower seems remote, while the rising power of China has already largely materialized. These local perceptions, too often forgotten by analysts due to the lack of sociological information on post-Soviet Central Asian and Afghan societies, are significant. They tap into the self-images that have an impact not only on public opinion but also, one way or another, on the long-term choices of political leaders.

In less than two decades, the geopolitical readings of Central Asia have multiplied: the southern margins of the former Russian Empire, the eastern pole of Washington's "Greater Middle East," the new "Far West" of China, the Caspian Sea as a historical place of conflict between Russia and Iran, a "Central Eurasia" where Slavic, Turkic, Persian, and Chinese cultures meet. These familiar interpretations invite neighboring and more distant states to project power in the region. However, power projection and mechanisms of leverage and implementation are two different things. Although the image of Central Asia as a land of new global confrontation between rising powers such as India and China may capture the imagination, sobriety should drive the analysis; Russia, the United States, and the European Union are all equally important there. And far from the glorifications of the geopolitical "crossroads of the world," the moves of Chinese and Indian actors remain marked by hesitation and, above all, pragmatic choices.

The revival of the so-called Great Game must be nuanced. First of all, the Central Asian states are not mere pawns, subject to competition between powers. They are independent actors that have a narrow margin to maneuver against their Russian, Chinese, and Indian neighbors but are still independent in their foreign policy decisions. Each of them has a very specific identity and divergent visions of its geopolitical environment. One does not regard China in the same way as Kyrgyzstan and Turkmenistan, and India as Kazakhstan or Tajikistan. Then, there is no binary opposition between major powers in Central Asia. Russia and the United States have not only conflicting economic interests, but also complementary ones in security. Russia and China appear to share control over the Central Asian regimes but will likely compete in the coming decades. China and India have common visions for the stabilization of Afghanistan, but mostly growing differences in the analysis of their interests in the post-Soviet Central Asia. In addition, other international players are present, mainly the European Union, Turkey, and Iran, but also Japan, South Korea, and the United Arab Emirates, among others.

One cannot think of Central Asia merely as a region of conflict between great powers, because it is also a space of complements and negotiation. In addition, despite its growing importance, Central Asia remains a peripheral place in many ways and has proved central only in security terms. For Russia, any destabilization of the area would have immediate impact on its own domestic security. For China, the implications aim directly at the stability in Xinjiang, and for India, in Kashmir. However, economically, Russia looks primarily to Europe and possibly the Far East, Beijing will continue to direct its gaze toward its economic partnership with the United States and the assertion of power in Asia, and Delhi will focus on its complex relationship with its neighbor Pakistan and on its growing economic relations with the United States and the European Union. The overvaluation of security in Central Asia contrasts with its economic role, which is more modest.

As for all the neighbors of the former Soviet Union, the disintegration of the country, the change of regime, and the introduction of a market economy have brought both benefits and risks: benefits via political partnerships and economic ventures, risks in terms of new geopolitical tensions and competition for the control of wealth. Since 9/11, the global "war against terrorism" launched by Washington has intensified security-driven views of Central Asia. The region is indeed subject to destabilization from Afghanistan, mainly through drug trafficking, which fuels the criminalization of the economy and state structures and finances clandestine groups claiming allegiance to Islamism. However, the long-term issues may be primarily economic. Indeed, Central Asia will be resistant to possible destabilization by betting on development, and this cannot be achieved without the involvement of neighboring powers. In Afghanistan too, the legitimacy of the central government will only be built on evidence of economic performance that will change the lives of its citizens. In this area of aid to Kabul, New Delhi is well positioned vis-à-vis Beijing, which has not had the humanitarian experience of India and is interested in Afghanistan because of its commodity market. Conversely, in aid to post-Soviet Central Asia, China heavily dominates India.

For China, the primary objective of its relations with independent Central Asia was to secure its borders with Kazakhstan, Kyrgyzstan, and Tajikistan— which it did by obtaining treaties demarcating borders, thus ending decades of conflict with the Soviet Union—and to prevent the region from becoming a rear base for Uyghur independence movements. Both objectives were achieved, although the latter can always shift in coming years. The security component is important in the context of the Shanghai Cooperation Organization, even if Beijing cannot eliminate Russian strategic supremacy there and has refocused on economic issues. For India, the establishment of relations with Central Asia did not have to go through a phase of border dispute settlements. New Delhi first analyzed its relation to the new states through the prism of its conflict with Pakistan: it has sought to halt Islamabad and to prevent Central Asia from offering Pakistan the famous "strategic depth" it lacks. Although post-Soviet Central Asia is not linked to the Kashmiri conflict, this is not the case with Afghanistan, which directly affects domestic Indian interests. The terrorist attacks in Mumbai in November 2008 may have been linked to the progress of the Pakistani army in the Taliban-controlled areas of northern Pakistan. Al-Qaeda losing power in the Af-Pak region correlates with new attempts to destabilize Kashmir. For India, the Afghan lens focused on Central Asia is thus central, whereas it is less important for Beijing. It was necessary to wait until around 2005 before China sought to involve itself in Afghanistan and coordinate its policies in Central Asia, particularly in Tajikistan, with those established in Kabul.

Although both countries want the settlement of the Afghan issue and stability in Central Asia, they differ on many levels in their reading of the global geopolitical environment. On one side, China is deeply concerned about U.S. presence in Central Asia and Afghanistan because it could reduce its room to maneuver in the region over the long term and even more in its settlement of the Uyghur and Tibetan issues. For its part, Delhi did not see any major disadvantages in U.S. presence in the middle of the continent and sought instead strategic rapprochement with Washington. On the other side, China has developed a modus vivendi with Russia in Central Asia, leaving Moscow with the impression of control in the region, while India has lost status with the Kremlin and is hardly close to regaining it. China sees Central Asia as a means to access the Iranian-Turkish Middle East, while Delhi frames the situation primarily in terms of Sino-Pakistani encirclement. Finally, China benefits from a multilateral instrument, the Shanghai Cooperation Organization, while India, though an associate member since 2005, has no regional platform to better structure its interest in the region. But Beijing and Delhi agree on one point: boosting Iran's presence in regional affairs and stopping its economic marginalization, which penalizes all neighboring states in terms of freight and hydrocarbon exports.

Despite the gradual normalization of Sino-Indian relations in recent years and the growth of bilateral trade, both countries are constrained in their possible cooperation by unresolved border problems, the sensitive issue of the Dalai Lama, and competition for geopolitical and economic leadership in Asia. The Sino-Pakistan partnership is not likely to disappear in coming years. Despite Chinese concerns about Islamist risks, the encirclement of India and access to southern seas via the port of Gwadar invite Beijing to respect Islamabad's sensibilities. The Chinese authorities are also concerned about Indo-U.S. rapprochement, especially regarding the military and nuclear weapons. In Central Asia, the two Asian powers are already competing in terms of access to oil, uranium, and other mineral resources, although the Indian business presence remains minimal compared to that of China. However, both countries are expected to be soon in competition in the service sector and communication technologies. Although for now Central Asia needs the Chinese "world's workshop," in the future it will focus more on its Indian "world's back office" neighbor.

One cannot be satisfied only with a geopolitical and economic reading of the rise of China and India in Central Asia: the questions are partly internal for the two countries and shaped by their apprehensions of cultural and religious issues. For Beijing and Delhi, there is a fairly direct link between their perceptions of Central Asia and their management of Islam. The Chinese and Indian Muslim minorities form internal otherness, structuring the political experience of both countries, although India has historically had a much less tense relation to Islam than China. In both countries, the independence of Central Asia has awakened ancient historical myths such as the Silk Road—a route through which the riches of Asia reached the Mediterranean by caravans, with a unique command of the extreme geographical conditions crossed. These cultural spaces now seen as separate—Central Asia, Afghanistan, Kashmir, and Xinjiang—have indeed been linked for centuries, through trade routes that drove the intense circulation not only of goods, but also of people and philosophical, religious (Buddhism and Islam), and architectural ideas.

The romantic evocation of the continental routes in a contemporary context, when three quarters of world trade is done by sea, brings up necessary improvements in areas forgotten by the large waves of modernization of the second half of the twentieth century. Xinjiang in China and Kashmir in India, besides being unstable and culturally and religiously different from their metropole, are economically peripheral and on the margins of the internal geopolitics in each of their two states. However, these weaknesses can be interpreted as potential forces. China-Central Asia and India-Central Asia relations must, therefore, be understood not only as the domain of international relations but also as an element of the domestic Indian and Chinese policy toward Xinjiang and Kashmir. These complexities make it difficult to separate rhetoric from reality; in the analysis of the actors themselves, historical myths and national narratives are fully associated with the redevelopment of continental routes. Over the long-term, covering the entire twenty-first century, the geopolitical orientation of Beijing toward the West and of Delhi toward the North is likely to evolve, but for now the second remains modest and reserved in practice.

The first part of this book examines the geopolitical projections of power in Central Asia. Moscow, Delhi, and Beijing interpret in their own ways both domestic and external security issues and project their knowledge on Central Asia depending on a specific historical and contemporary background (Marlène Laruelle, Emilian Kavalski, and Jean-Pierre Cabestan) and, for China and India, on complex relationships to Afghanistan and Pakistan (Meena Singh Roy, Swaran Singh). The second part of the volume discusses Central Asia as a space to analyze cooperation, parallelism, and competition between India and China (Jean-François Huchet, Zhao Huasheng, Basudeb Chaudhuri, and Manpreet Sethi). In a third part, the book delves into the economic realities of Indian and Chinese implementation in Central Asia: participation in the exploitation of Caspian resources (P. L. Dash), the divergent economic presence of the two players and their specific market niches (Sébastien Peyrouse), and the issue of reconstruction in Afghanistan (Gushan Sachdeva). In a fourth step, we focus on revisited historical backgrounds and disputed religious modernities in Central Asia, India, and China: reworking of historical memories and myths related to the Central Asian space in India (Laurent Gayer), the issue of Uyghur Islam in Xinjiang and its links with the post-Soviet world (Rémi Castets), the growing influence of Indo-Pakistani religious movements in Central Asia (Bayram Balci).

PART I

Negotiating Projections of Power in Central Asia

Russia Facing China and India in Central Asia: Cooperation, Competition, and Hesitations

MARLÈNE LARUELLE[1]

If Russia reacts strongly to the presence of other international actors in its former Central Asian "backyard," this reaction, although based on objective economic competition, is mainly due to subjective perceptions related to balance of power issues. Russian global geopolitical interests have substantially changed since the end of the cold war and the Kremlin is still in the process of adjusting its perceptions of the international scene, with difficulties in identifying its long-term partners and enemies.[2] Moscow is regularly concerned with the United States' advances in Central Asia, and sometimes with those of the European Union, comments with disdain on Turkey's presence in the 1990s and on that of the United Arab Emirates since 2000, has relatively little worry about the activities of Iran and India, and dare not take a critical position on the Chinese presence. The Sino-Russian partnership was strained compared to the good relations between Moscow and New Delhi, with the Treaty of Friendship and Cooperation signed between the two countries in 1971 in the midst of the Sino-Soviet conflict. However, since the 1990s, Russia has played the rapprochement card with China, forging a new ally in a world set to become multi-polar. The relationship between Moscow and New Delhi has suffered from Russia's ventures toward China, but the doubts the Russian elite harbor regarding Beijing's ulterior motives in the Far East and Central Asia have not been erased—quite the contrary.

Since the second half of the 1990s, Moscow has revived the idea of a Russia-China-India triangle, symbolized by the personality of Evgeni Primakov and then Vladimir Putin, who sought to realize it through strategic and economic partnerships.[3] The three countries, alongside Brazil, were grouped by economists under the acronym BRIC, four high-growth countries whose weight in the global economy is on the rise. Despite its first summit in Russia in June 2009, the BRIC has no specific activities, and even the trio Russia-China-India does not actually follow a joint policy.[4] China and India have complex relations tinged with competition; Russia and China are officially strategic partners but reticence and mistrust exists in spades on both sides,[5] whereas Russia and India are struggling to boost their potential alliance. Moscow views the world through a prism of fear of its confinement to the periphery of international decision-making.[6] Its understanding of Beijing and New Delhi's role in Central Asia should be analyzed according to this difficult power projection, in which the two Asian states are mirroring a forthcoming Russian decline.

The Paradox of Russian Interests in Central Asia

Russia is a power unlike others in Central Asia, insofar as it is the region's former colonizer—a role that started in the nineteenth century and even in the eighteenth for some of the northern parts of Kazakhstan. This legacy has its positive and negative aspects. It has been positive insofar as it involved a long period of Russian-Central Asian cohabitation that gave rise to common feelings of belonging to the same "civilization." It has been negative insofar as it has accrued all the political resentment and cultural misinterpretations of the colonizer-colonized relationship. Russian-Central Asian relations are, therefore, complex, with both actors having a highly emotional perception of its relation to the other. After the disintegration of the Soviet Union, Russia's standing as the former colonial center presented it with many difficulties; holding Moscow at bay was a top priority. Resounding critiques rang about "Russian colonialism," but they lasted only for a brief period.[7] A period of less than two decades has, therefore, sufficed for this common legacy to be positively reshaped and for Moscow to succeed in inverting the Soviet past and turning it into an asset of shared proximity. Since 2000, Russia has once again become a respected power in Central Asia, where its economic and geopolitical revival is widely admired.[8] During Vladimir Putin's two terms (2000–2008), it succeeded in returning to its status as the number one partner of the Central Asian states. Although Russia has succeeded rather well in its return to Central Asia, it is also in the process of becoming a power like the others in the region.

On the multilateral level, the two Moscow-initiated organizations the Eurasian Economy Community (created in 2000 on a Kazakh proposition) and the Collective Security Treaty Organization (founded in 2002) today function as the major institutional frameworks of Russian-Central Asian cooperation. On the bilateral level, Moscow is again a first-order political, strategic, and economic ally. The Kremlin has made a show of its abiding political support for the Central Asian regimes, a rapprochement facilitated by the common struggle against the so-called Islamist threat. From 2000 on, Russia has once again become the primary political point of reference for the Central Asian regimes, which are attracted neither to Western parliamentary systems nor to the Chinese single-party order. The Central Asian leaders did not wait for the Putin experiment to limit political expression and the autonomy of "civil society," but they have been able to draw additional legitimacy arguments from the Russian example. This political rapprochement has had a significant impact on Central Asian societies: political reforms for democratization have been impeded, the activities of NGOs and "civil society" are increasingly curtailed, and gaining access to new technologies and media like the Internet made more difficult.[9]

On the cultural level, Russia clearly has an advantage. Russian is still the most-spoken international language in the region with status in three states: the official second language in Kyrgyzstan and a language of interethnic communication in Kazakhstan and Tajikistan.[10] Growing labor migration reinforces the weight of Russia, which found a new linguistic and cultural influence in the region.[11] But Moscow's ability to seduce Central Asia remains limited. Its recognition of South Ossetia's and Abkhazia's independence in the summer of 2008 angered Central Asian authorities, which call for preserving the borders of the former Soviet Union and limiting the propagation of secessionist movements.

Moscow's three main practical concerns in the region are security issues, political influence, and economic presence (especially in gas and oil exports). In the economic domain, Russia has regained a dominant position since 2000 but has lately lost several important battles. Russian-Central Asian trade bounced back around 2000 and tripled between 2003 and 2007, shooting from US$7 billion to US$21 billion, a third of which is from the hydrocarbon sector.[12] Since 2006, Russia regained status as the primary commercial partner of Uzbekistan (imports and exports) and Tajikistan (but it is actually second to China in imports and to the European Union in exports) and as the second-largest commercial partner of Kazakhstan and Kyrgyzstan (it is the largest importer into the former and the largest exporter into the latter) but has only risen to the rank of fifth partner for Turkmenistan.[13] In the Central Asian trade sector, Russia will in all likelihood be overtaken by China in only a few years, if this is not already the case in Kazakhstan and Kyrgyzstan.[14]

Russia remains a dominant economic actor in Central Asia if energy is taken into account. It is a significant one in heavy industry and infrastructure, both of which are old Soviet specializations. Yet it is a relatively modest and rather uncompetitive player in terms of small and medium-sized enterprises and new technologies. This stratification offers a more general reflection of the Russian economy as a whole, which is still in a rent-seeking logic and is having problems diversifying. But it is also explained by the state of the Central Asian economies, in which small and medium-sized enterprises and new technologies struggle to find their place. These economies are destined to serve, above all, as transit zones for Sino-European trade, hence the emphasis on infrastructure and freight-related services.

One of the key domains of the Russian presence in Central Asia is that of regional security. Since the early 1990s, it has been the primary driving force behind Moscow's continued presence in the region; however, since 2000, the mechanisms of this collaboration have been profoundly transformed. Russian security challenges in Central Asia are multiple. Any destabilization in the weakest states (Kyrgyzstan and Tajikistan) or the most unstable (Uzbekistan) ones will have immediate repercussions in Russia: Islamist infiltration in the Volga-Ural region and the North Caucasus, indeed in the whole country; an increase in the inflow of drugs reaching the Russian population, which is already a large target of drug traffickers; the loss of control over the export networks of hydrocarbons, uranium deposits, strategic sites in the military-industrial complex, and electricity power stations; a drop in trade exchanges; the loss of direct access to Afghanistan; and an uncontrollable surge of migrants, in particular refugees. For Moscow, the securing of the southern borders of Central Asia is seen as a question of domestic security—not by "imperialism," but by pragmatism. The 7,000 kilometers of Russian-Kazakh border in the heart of the steppe are nearly impossible to secure. They require clandestine flows to be better controlled downstream, as it were, which goes to confirm Central Asia's role as a buffer zone for Russia.[15]

Any consideration of Russia's successes in Central Asia proves complex. Indeed, despite its return since 2000, Moscow has truly lost its stranglehold over the region. And yet the Kremlin itself has never envisaged returning to a Soviet-style situation, nor has it tried to reintegrate the Central Asian states politically by including them in the Russian Federation. Though Moscow hopes to remain Central Asia's number one partner, it no longer imagines its presence to be exclusive. The Kremlin has, therefore, learned, to its detriment early on, to compromise with other international actors, as evidenced by Vladimir Putin's post-September 11 acceptance of the opening of American bases in Uzbekistan and Kyrgyzstan and cooperation with China in the framework of the Shanghai Cooperation Organization. Russian elites are pragmatic. They know that the Central Asian states are prepared to exploit the factor of international competition in their own national interests and that they will not give Russia any preferential treatment on the grounds of sympathy alone. The conclusion, therefore, seems to be that Russia's return has been a partial success insofar as Moscow has again become an important partner and a legitimate ally in a Central Asian market that is no longer monopolistic. However, its positions are by no means guaranteed and remain subject to global geopolitical hazards.

It is true that neighboring regional powers, such as India, Iran, and Turkey, lack the means to dethrone Russian supremacy; however, this is not the case with China, which is going to prove problematic for Russia over the long term.

The Russian presence in Central Asia is also dependent on economic stakes. Since 2000, oil and gas–related income has provided Russia with leverage that it did not previously enjoy—but one that it could lose if the global financial crisis, combined with falling world hydrocarbon prices, were to continue. Russia's capacity to invest in Central Asia might then run into difficulties and this would have a direct impact on its political influence. Moreover, Russia's weight in Central Asia depends not only on global geopolitical and financial redistributions but also on domestic factors. Central Asia is conceived of as an intrinsic and natural part of the Russian sphere of influence; political submission and economic control are desired, but not cultural proximity since it provokes anxiety. In fact, in Russian public opinion, Central Asia is usually associated with notions of Islamism, terrorism, and the mafia, while positive references emphasizing the historical and cultural ties to the Central Asian peoples are extremely rare. This generalized disdain for Central Asia provides the negative context in which the Russian intellectual and political elites conceive of the utility of the region in terms that are not only economic and geopolitical but also social.[16] This situation can partly be explained by Moscow's general view of Central Asia as part of its de jure sphere of influence in Eurasia. For the moment, therefore, Russia has almost no long-term vision of the relations it would like to entertain with its "South," nor any strategy to propose that would offer Central Asia any status other than as Moscow's geographical and political appendix. The Kremlin is still inclined to think of Central Asia as an acquired zone of influence and the Central Asian governments feel this is disrespectful.

Russia does have arguments working in its favor of which it might not be aware. The young generation of Central Asian elites, particularly in Kazakhstan and Kyrgyzstan, who were partly educated outside of the CIS, is likely to have a more critical view of Russia; however, this does not necessarily entail that Russophobic circles are about to occupy positions of power. On the contrary, the pacification of memory relating to the Soviet past—the idea that the Soviet Union was instrumental in achieving independence, that Russia continues to be the path to Europe, and that there is a specific "Soviet" or "Eurasian" civilization—includes pro-Russian arguments that bear much weight among the Central Asian upper and middle classes. However, to date, the Kremlin has not developed any coherent or expansive cultural diplomacy for the post-Soviet space. Indeed, it has no real suggestions to offer about how to sustain the role of the Russian language and Russian-speaking culture in Central Asia, although this is now starting to become an important question in decision-making circles. Russia could conserve its key role in Central Asia were it to give itself the means to do so. These would include complying with the Central Asian states' wishes for investments in economic sectors other than those of hydrocarbons, continuing to train Central Asian political and military cadres, promoting the Russian language among local populations, creating a more secure working environment for the millions of Central Asian migrants settled in Russia, reviving the intellectual exchanges between Russia and Central Asia (which have practically ceased), and getting involved in the policies of lobby-formation and building influence networks among the young generation that stands to inherit the reins of power.

Russian Views on China and India

Since the beginning of the 1990s, Russia and China have sought international rapprochement. Territorial conflicts from the Soviet period have been settled, even though traces remain in memory and ideological differences have been obliterated while political, economic, and strategic relations have solidified.[17] A 2005 treaty definitely resolved the land issues; a 2008 border agreement complemented it. The contemporary Russian-Chinese alliance rests in large part on their mutual desire to oppose the would-be hegemony of the United States and to defend a multi-polar organization of the world. The two capitals follow similar interpretations of the Chechen question in Russia and the Tibetan and Uyghur issues in China, although Moscow's recognition of the independence of South Ossetia and Abkhazia in August 2008 caused confusion among the Chinese authorities. Moreover, this Sino-Russian "anti-Western" partnership remains ambivalent. Neither of the two regimes wants a direct confrontation with Washington or a slowdown of their integration into the international community. The alliance game between the two countries is, therefore, lacking a "values base" and could be tested by a geopolitical change for either actor. The Sino-Russian partnership is, in fact, based on the tacit recognition by China of its need for Russian military and geopolitical assistance, but Russian policymakers are increasingly concerned at the speed with which China is narrowing the strategic gap.[18]

The balance of power will lean toward China in coming decades, reflecting in a zero-sum equation Russia's decline as a strategic and economic superpower. Moscow feels threatened by the economic dynamism of China, worried about China's reluctance to grant it status as an Asian power, and apprehensive about being put aside in the new Asian security architecture, particularly on the Korean issue. The Russian strategic documents do not openly mention threats from China and continue to identify the West and NATO as the primary dangers, along with terrorism, because they reflect a near-term reading of the issues. However, long-term Russian projections cannot avoid the "China question."[19] Moscow must, therefore, cultivate ambiguity toward

Beijing and show its confidence in a peaceful Chinese future, all the while seeking counterweight from Japan, India, and Europe. On the economic level, Russia wants to develop its partnership with China, particularly in the energy domain, in order to have a counterweight to its tense relations with Europe. Bilateral trade reached US$56 billion in 2008, but only US$40 billion in 2009 because of the economic crisis,[20] and is set to double sometime around 2012 to 2015. However, because of shuttle trade in the Far East and arms sales, the real scale of Russo-Chinese trade is much higher than official statistics from Russian and even Chinese sources suggest. Trade figures clearly show Russia's orientation toward primary resources and China's role as a supplier of finished products; the level of cooperation between the two countries in manufacturing and investment is almost negligible.

Military and strategic cooperation constitutes a central element of the Sino-Russian partnership, but this statement must also be qualified. Real cooperation in the military domain remains modest. Joint military exercises, organized for the first time in a bilateral manner in 2005 and thereafter in a multilateral way under the auspices of the Shanghai Cooperation Organization, ought not to cause any illusions. They are above all designed to demonstrate the military power of the two countries, to send signals to the United States, and to secure Central Asia; neither of their military leaders envisages any exchange of strategic information or a deeper partnership.[21] Arms exports and military technology transfers, one of the strong points of the Moscow-Beijing relationship, are also full of paradoxes. Since the beginning of the 1990s Russia has supplied 85 percent of Chinese arms imports, while these large Chinese weapons contracts represented about 40 to 50 percent of Russian arms exports for the period 1992–2004.[22] Chinese orders are, therefore, vital to the Russian military-industrial complex. However, there has been a quick drop in Chinese orders—in 2009, China's share in Russian arms exports fell by 18 percent[23]—forcing Russia to turn to new markets in South Asia, the Muslim world, and Latin America. Tensions are emerging, since Beijing is looking to increase its technology purchases, filling Russia with fear that it will lose its Chinese client, which will perhaps soon no longer need Russian expertise and know-how. In coming years, China is expected to become autonomous from its Russian military tutor and could even become a competitor on the international arms markets.

Though Chinese elites have quite a positive vision of Russia, Russian elites continue to be distrustful of the former historical enemy and are pointedly Western-centric. The Chinese model of development appeals to only a small portion of the Russian political class.[24] The relation to China presents equally major problems for the internal balance of the Russian Federation. The stakes are no longer formulated in the political terms of secessionism; nonetheless, Russia continues to confront centrifugal tendencies at the economic level.[25] The Russian authorities are, therefore, increasingly disquieted by China's

economic dynamism, which will enable Beijing to present itself as the dominant power both in Central Asia and the Russian Far East within the next
decade.[26] As was manifest in Russia's hesitations regarding the route of future
gas and oil pipelines in Siberia, Moscow is banking on Beijing's rivalry with
Japan and South Korea to avoid having to go head-to-head with China, which
would certainly not be to its advantage. The population deficit in Siberia also
elicits numerous fears linked to the risk of the disappearance of ethnic identity
and threats to territorial integrity.[27] Over the long term, points of friction
seem likely to increase. These will combine rivalry over Central Asia and the
North Pacific as a sphere of influence with the tensions in the Russian Far
East that arise from the strong demographic and economic pressures that pull
this region into Beijing's sphere of influence.[28]

While the Sino-Soviet conflict undermined relations between Moscow
and Beijing, India's historical position of nonalignment has always been well
received by the Russian elite. The two countries have never experienced epi-
sodes of conflict in their bilateral relations and India is not seen as a long-term
threat. Today, both countries share a vision of international balances—that
include resistance to U.S. hegemony, denial of the use of "double standards"
by the West, the democratization of international relations, state sovereignty,
and right to pursue specific models of development—nonetheless, the recent
strengthening of Indo-American partnership could question this common
trend.[29] The two countries support each other in their international ambitions. Moscow displays an understanding perspective on the Indian nuclear
issue, while asserting it could not amend the NPT for New Delhi, and supports India's candidacy for a permanent seat in the United Nations Security
Council and full membership to the Shanghai Cooperation Organization. For
its part, India supports Moscow's bid for the Asia-Europe Meeting (ASEM)
and endorses Russian ambitions to be more involved in the political and economic life of Asia. In addition, both countries are strengthening their relations
with the Muslim world and share a concern about terrorism, backing Russia's
management of the situation in the North Caucasus and India's in Kashmir.
Russia has also held back its relations with Pakistan in order to satisfy its Indian
ally, even though Moscow has tried for several years to initiate a new dialogue
with Islamabad.[30] Finally, the rise of China similarly preoccupies Moscow and
New Delhi, although both have opted for a pragmatic good neighbor policy.

Despite a very favorable political background, Indian-Russian economic
relations have struggled to take off. While the Soviet Union had been the
largest trading partner of India, commerce between the two comprises today
no more than 1 percent of their total trade. During the 1990s bilateral relations collapsed under the weight of Russian difficulties, and the Indian elite
complained about the contemptuous and Westward attitude of Yeltsin's governments.[31] Indian businessmen also protest about the Russian investment

climate and the difficulty in penetrating opaque markets. However, from 2000, Vladimir Putin boosted the Indo-Russian relationship and sought to turn into reality Primakov's discourse on the Russian-Indian-Chinese triangle. A strategic partnership was signed the same year. In 2005, trade relations returned to their 1990 level, about US$5 billion.[32] A roadmap signed in 2006 foresaw US$10 billion in trade in 2010 and US$20 billion in 2015. However, the Comprehensive Economic Cooperation Agreement (CECA) that is expected to enhance bilateral relations depends on Russia's accession to the World Trade Organization, which is still delayed. The global crisis of 2008 and 2009 has also sharply curtailed the economic expectations of both partners.

On trade, existing exchanges are small compared to potential ones. The North-South corridor projects in Central Asia and Iran are not effective. The amount of trade flowing through them remains low, while trade between the EU and India is expanding very quickly. Bilateral trade is mainly composed of raw materials exported from Russia to India, which in turn exports mainly tobacco and tea, and of cooperation in energy infrastructure and basic metals. In terms of hydrocarbons, ONGC and Gazprom collaborate on joint projects at Sakhalin-I for the former (India's largest single foreign investment, at approximately US$3 billion)[33] and in the Bay of Bengal for the latter.[34] Three areas of cooperation seem particularly promising: diamonds, since Russia is the largest producer of rough diamonds in the world and India its primary processing center; new technologies, since Russia is seeking to diversify its economy and catch up in IT and nanotechnology, an area where India is ahead; and space, as India seeks to assert its status as a space power and will need the know-how that Russia can offer (Roskosmos has already several joint projects with the Indian Space Research Organization). Finally, nuclear cooperation, long blocked by India's refusal to accede to the NPT, will develop since New Delhi has come to an agreement with the International Atomic Energy Agency. In 2009, the two countries signed a cooperation treaty that provides for the supply of Russian uranium to the Tarapur plant in Maharashtra, the construction of several Indian reactors by Russian firms, and the commitment of Moscow to the Indian civilian nuclear program in general.[35]

In contrast to these still-modest commercial relations, the Indo-Russian partnership is particularly powerful in the military sector. Strategically, the two countries have been conducting biannual joint counterterrorism exercises since 2002 and have held naval exercises in the Indian Ocean in 2003. Above all, India remains one of the main foreign partners of the Russian military-industrial complex. Aviation and naval orders drive this bilateral military cooperation.[36] Two-thirds of the Indian army is still equipped with hardware produced in the Soviet Union or Russia; in the 1990s, it was one of Russia's only customers, ensuring the industry's survival against the deficit of domestic orders. As noted by Isabelle Facon, the specificity of Indo-Russian military relations lies in developing partnerships for the design and joint production of armaments.[37] Since Soviet times, Moscow has been ready to deliver to India performance systems that it refused to its Warsaw Pact allies, or to China today. New Delhi is a very demanding customer in terms of technology, asking for hybrid production that takes after European technological advances, pushing Russia to upgrade and internationalize. Moreover, New Delhi is seeking to support industrial projects, forcing Russia to sell its production licenses. Thus, if China remains a more significant customer of the Russian military-industrial complex in quantitative terms, India is positioning itself ahead of Beijing in qualitative terms. The impact of Chinese industrial capacity and the growing competition it poses in the armaments market worry Russia, but Indian capabilities, which are much smaller (paradoxically in the light of its success in knowledge industries), do not.[38] But none of these is a captive market. India is trying to diversify its military purchases from the United States, European Union, and Israel, which openly compete against Russian companies. Moscow, meanwhile, is working more with countries such as Vietnam, Algeria, and Venezuela than with India and China.

The Russia–China–India Triangle in Central Asia

Russian elites do not view the rise of China and India in Central Asia in the same way. The first is seen as a key player, for better or for worse, while the second is a potential one, welcome but still distant. Moscow does not yet understand India as a possible counterweight to China's presence: the imbalance of power between the two Asian states is too large, and Beijing is in any case not expected to fade from the Central Asian scene. The real issue for Moscow is, therefore, how to manage Beijing's inevitable competition without completely losing control of Central Asia, as India is not expected to disrupt this duo for several decades. The perception of Beijing as a powerful and ambitious competitor in Central Asia is a recent phenomenon in Russia and remains difficult to analyze, since diplomatic relations between the two countries are fraternal. Russian official publications remain silent on the subject, and it is only off the record that experts dare to raise the issue of Chinese potential to dethrone Russian dominance in the region.

The current Sino-Russian condominium in Central Asia must be qualified. It is based not only on certain economic and geopolitical realities, but also on several unstated issues. On the geopolitical level, Beijing and Moscow share the same view of the dangers that they face in Central Asia, a potentially unstable area with the risks of Islamism, state failure, and drug trafficking. Therefore, Moscow, like Beijing, gives support to local authoritarian regimes, arguing that they are the guarantors of stability and secularism to counter the Islamist risk. However, even if Beijing supports the fight against Islamism in Central Asia, its friendly relationship to Islamabad is problematic for Moscow, which understands it as a dangerous contradiction of China's foreign policy.[39] Moscow and Beijing also share a similar commitment to fight against Western influence in the region and U.S. containment of their territory through the establishment of friendly regimes via "color revolutions." On the strategic level, Beijing wishes to preserve Russian domination: it prefers to let Russia pay the heavy costs of military security and guarantee the survival of unstable regimes.[40] Indeed Moscow's military presence in the region does not raise any problems for China, which needs Russian support both to nip its own separatist movements in the bud and to act as a check on Western influence and growing competition from Washington. However, Moscow's policy toward China exhibits double standards, since it is preventing China from gaining a foothold in the CSTO's military. It is, therefore, unlikely that a real partnership could develop between the SCO and the CSTO: Moscow has no interest at all to merge an efficient organization over which it has total control with the SCO. This situation is not expected to change fundamentally in the years to come. As long as China agrees to give Russia its supremacy in terms of political and strategic influence, the interests of Russia and China in Central Asia are complementary.[41]

On the economic level, the sense of rivalry grows. For the time being, both powers seem to be accomplishing their objectives without producing any direct confrontations, but this situation could change in the future. China is undergoing a period of exponential growth and devouring primary resources, while Russia is using its economic revival to specialize in primary resources and heavy industry. Central Asia is, therefore, going to be an important component in the economic strategy of both its neighbors. China's growing presence is liable to run into direct competition with Moscow's intentions in the region's gas sector, where the interests of Chinese companies are in direct conflict with those of Gazprom; in Kazakhstan's oil sector, where Lukoil and CNPC have already squared off; in gaining access to Uzbek and Kazakh uranium, a key component of the nuclear power programs of both Russia and China; in the mining of other precious minerals; and in the increasingly coveted export of electricity, which must go to either Russian or Asian markets. To the contrary, the development of Central Asia, especially Kazakhstan, as a transit route for Chinese goods to Europe via Russia will remain an important element of the economic partnership between China and Russia in the region. Whether Russia wants it or not, Beijing seems destined over the medium term to dominate the Central Asian market in many sectors, thanks in particular to its financial and banking clout, which Moscow lacks.[42]

Furthermore, the Sino-Russian alliance could be undermined by developments occurring outside Central Asia: renewed tensions in the Far East due to the growing weight of China and the dangers it poses for the territorial unity as well as the political and economic development of the Russian Federation (with an overly assertive China taking ascendancy over Russia); the removal of their common geopolitical interests on the international level. The Kremlin knows that Beijing's growing international importance will weigh against it should the two states' interests diverge. Moreover, the historical rifts between the Russian Empire and the Middle Kingdom have not vanished from people's minds, neither has the fact that these two worlds will become rivals for influence in the coming years when China starts to assert itself as a cultural power. If the Chinese authorities were to consider, for whatever reasons, that they ought to modify their activities in Central Asia and involve themselves in political issues, not just in economic ones, then Chinese interests would come into conflict with those of Moscow. In this case Central Asia would find itself trapped in the middle, and the required choice between Moscow and Beijing would exacerbate tensions among Central Asian elites.

Unlike China, India rises as a less ambiguous partner that is not soon expected to change the geopolitical balance in Central Asia in fundamental ways.[43] Seen from Moscow, the growing involvement of New Delhi is a positive element: it could potentially counterbalance the omnipresent China, but more concretely and immediately, it will halt the settlement of Pakistan in the region.[44] Moscow delegates New Delhi to the surveillance of Pakistani activities in Central Asia. Both countries, in fact, share similar concerns about Islamist terrorism, the Pakistani nuclear program, and possible destabilization of Islamabad, which would jeopardize attempts to stabilize Afghanistan.[45] However, the reconciliation New Delhi displayed with Washington could be problematic for Moscow if it were to transform India into a representative of U.S. interests in the region. From an economic standpoint, Russia and India are potentially competing for the export of hydrocarbons (the rallying of Delhi around the Turkmenistan-Afghanistan-Pakistan pipeline project is an example), electricity, and uranium. But India's actual capacity to leverage is currently too small for it to be perceived as contrary to Russian interests.

Strategically, Moscow favors the greater involvement of India in Central Asia. Russia has not objected to the fact that some Tajik, Turkmen, and Uzbek officers received training in Indian military academies in the 1990s, to the Indian-Uzbek bilateral working group to fight international terrorism (created in 2003 but having few assets), to India's training of specialists for the future Kazakh Caspian military fleet, to the joint Indian-Tajik exercises held in 2004 on the Fakhrabad polygon in the Khatlon region, or to the courses offered in aerospace engineering for Tajik helicopter pilots. Instead, Russia stands to gain from Indian rapprochement through deals struck with the Central Asian military-industrial complex—for example, New Delhi's purchase of aircraft from the Tashkent aviation factory TAPO and possible procurement of torpedoes or torpedo parts produced in Kyrgyzstan and Kazakhstan. Indian-Kazakh space negotiations also seem quite compatible with the Russian-Indian partnership in this area and could lead to tripartite space activities.[46]

However, the uncertainties surrounding the negotiations around the base of Aini, in Tajikistan, revealed ambiguities. In 2002, Delhi and Dushanbe signed a military technical agreement for the renovation of the Aini base, which had served in the Soviet era as a repair shop for helicopters and as a base of operations during the Soviet intervention in Afghanistan, before being closed in 1985.[47] India has invested nearly US$20 million to restore the runway and construct aircraft hangars and a control tower. The Indian military authorities sent a squadron of Kiran aviation trainers to educate Tajik pilots. About 150 Indian military personnel, mainly engineers and support staff, have been brought there. However, Delhi's hope was to transform Aini as its first military base outside its borders, by stationing Mi-17 helicopters and then Mig-29 fighters at the site.[48] In 2006, Russian and Indian diplomats opened talks on India's possible role within the Collective Security Treaty Organization and the sharing of the Aini base. The early negotiations seemed to favor an agreement, but in 2007, Moscow let it be known that it opposed Indian deployment in Tajikistan and asked the Tajik authorities to revoke Indian access to the base.[49] It appears that the Kremlin's refusal was not directly linked to any Central Asian issue; it was rather a desire to punish New Delhi for seeking to diversify its weapon sales partners. Military cooperation between Russia and India is not so strong that Moscow would agree to share its "backyard" in Central Asia. New Delhi's rapprochement with Washington being problematic for Russia, the Indian base in Tajikistan could in this view potentially turn into a base that could offer the United States indirect access to Central Asia.

Conclusions

The Indo-Russian relationship is not confrontational, but for various reasons. On the geopolitical front, the common objectives of both countries are expected to continue, but economically their alliance is primarily temporary. India is welcome in Central Asia if it limits the presence of Pakistan there; for Moscow, a non-Islamic power is preferable to any other, especially since the Russians themselves are heavily involved in the fight against radical Islam. The rise of India is also well regarded as a counterweight to Chinese influence. The more players there are, the more Beijing will need to negotiate with its partners, leaving it unable to act unilaterally and enabling the Kremlin to avoid confrontation with Beijing. However, New Delhi's tight rapprochement with Washington would introduce elements of distrust on the part of Moscow, moderating the Indian "utility" for Russia in Central Asia. Moreover, the Indo-Russian partnership is based on specific circumstances. Moscow wants New Delhi to be more involved in the region because Indian presence there is still limited. Economically, both countries could be in conflict in the decades to come, when southern markets come to compete in earnest with the northern ones, which is not the case for the moment. India is, therefore, seen positively by Moscow on two levels: geopolitically over the long-term in order to deal with Pakistan and China, but with an American downside; economically only over the mid-term because it is still too small a player in Central Asia for the Kremlin to be upset.

Russia's relationship with China is much more complex because the latter is positioned in relation to not only Central Asia but also the Far East. China thus arises as a bilateral problem for Moscow, where it awakens old identity fears (what Bobo Lo rightly called the "Mongol syndrome"[50]), raises concerns about the territorial unity of the country in the future, and demands a fundamental rethinking of the geographic depth of the Russian Federation and the status of its Pacific flank. Central Asia is, therefore, a factor among others in the Sino-Russian pairing, which lacks positive values and is based on a negative geopolitical alliance against the United States. In Central Asia, the condominium between Moscow and Beijing works well but could lack the resilience to resist political, strategic, or economic downturns. It is in essence based on short-sighted geopolitical agreements—support for the regimes in power and rhetorical appeals for stability—and the ability of China to curb its ambitions in the region, recognizing Russia's military and strategic supremacy and positioning itself as a faithful second. However, economic competition is already palpable and will become increasingly important in the years to come. Any change in the overall balance between Moscow and Beijing may upset the Kremlin, which is unprepared to accept China as a top-notch political or cultural power. The Sino-Russian "axis of convenience"[51] is all the more paradoxical as both Russian and Chinese elites are really geared toward the West, but geopolitical and economic imperatives call them to order in Central Asia.

Notes

1. Senior Research Fellow, Central Asia-Caucasus Institute and Silk Road Studies Program, SAIS, Johns Hopkins University, Washington, DC.

2. A. Tsygankov, *Russia's Foreign Policy: Change and Continuity in National Identity* (Lanham, MD: Rowman & Littlefield, 2006); D. Averre, "Russian Foreign Policy and the Global Political Environment," *Problems of Post-Communism*, vol. 55, no. 5, 2008, pp. 28–39.

3. E. Wilson Rowe, S. Torjesen, *The Multilateral Dimension in Russian Foreign Policy* (London: Routledge, 2008).

4. Nivedita Das Kundu, "Russia-India-China: Prospects for Trilateral Cooperation," *Aleksanteri Papers*, 2003, <http://www.helsinki.fi/aleksanteri/english/publications/contents/ap_3–2004.pdf> (accessed February 10, 2010).

5. B. Lo, *Axis of Convenience. Moscow, Beijing, and the New Geopolitics* (Washington, London: Brookings Institution Press and Chatham House, 2008).

6. On this topic, see the demonstration proposed by D. Trenin, *Integratsiia i identichnost'. Rossiia kak "novyi zapad"* (Moscow: Evropa, 2006) or, in English, *Getting Russia Right* (Washington, DC: Carnegie Endowment for International Peace, 2007).

7. K. Syroezhkin, "Central Asia between the Gravitational Poles of Russia and China," in B. Rumed (ed.), *Central Asia. A Gathering Storm?* (Armonk: M. E. Sharpe, 2002), pp. 169–207.

8. J. Perovic, "From Disengagement to Active Economic Competition: Russia's Return to the South Caucasus and Central Asia," *Demokratizatsiya*, no. 1, 2005, pp. 61–85.

9. M. Emerson, J. Boonstra, N. Hasanova, S. Peyrouse, M. Laruelle, *Into EurAsia—Monitoring the EU's Central Asia Strategy* (Brussels-Madrid: CEPS-FRIDE, 2010).

10. In the latter, the status of Russia has been questioned in an increasingly open manner since the end of 2009.

11. M. Laruelle (ed.), *Dynamiques migratoires et changements sociétaux en Asie centrale* (Paris: Petra, 2010).

12. V. Paramonov, A. Strokov, *Ekonomicheskoe prisutstvie Rossii i Kitaia v Tsentral'noi Azii* (Defence Academy of the United Kingdom, Central Asia Series, 2007).

13. See European Union statistics, <http://ec.europa.eu/trade/trade-statistics>, October 2009 (accessed February 10, 2010).

14. For more details, see S. Peyrouse, "The Economic Aspects of the Chinese-Central-Asia Rapprochement," *Silk Road Papers*, Washington, DC: The Central Asia-Caucasus Institute, 2007.

15. M. Laruelle, "Russia in Central Asia: Old History, New Challenges?" *EUCAM Policy Papers*, September 2009.

16. M. Laruelle, "Russian Policy on Central Asia and the Role of Russian Nationalism," *Silk Road Papers* (Washington, DC: The Central Asia-Caucasus Institute, 2008).

17. E. Wishnick, *Mending Fences with China: The Evolution of Moscow's China Policy, 1969–99* (Seattle: University of Washington Press, 2001).

18. V. Paramonov, A. Strokov, *Russian-Chinese Relations: Past, Present, Future* (Defence Academy of the United Kingdom, Russian series, September 2006); B. Lo, "China and Russia. Common Interests, Contrasting Perceptions," *CLSA Asian Geopolitics* (London: Chatham House, May 2006).

19. V. Shlapentokh, "China in the Russian Mind Today: Ambivalence and Defeatism," *Europe Asia Studies*, vol. 59, no. 1, 2007, pp. 1–22.

20. S. Blagov, "Russia Moves into Trade Surplus with China," *Asia Times*, February 18, 2010, <http://www.atimes.com/atimes/Central_Asia/LB18Ag01.html> (accessed February 20, 2010).

21. J. P. Cabestan, S. Colin, I. Facon, M. Meidan, *La Chine et la Russie: Entre convergences et méfiance* (Paris: Unicomm, 2008).

22. I. Facon, "Les relations Chine-Russie vues de Moscou: Variable de la politique étrangère ou réel partenariat stratégique?" *Les Cahiers de Mars,* no. 183, 2005, pp. 100–110; and I. Facon, "The Modernisation of the Russian Military: The Ambitions and Ambiguities of Vladimir Putin," *Russian Series* (Conflict Studies Research Centre, Royal Military Academy, Sandhurst, 2005).

23. S. Blagov, "Russia Moves into Trade Surplus with China."

24. A. Lukin, *Medved' nabliudaet na drakonom. Obraz Kitaia v Rossii v XVII–XXI vekakh* (Moscow: AST Vostok-Zapad, 2007); a previous version in English is available, *The Bear Watches the Dragon: Russia's Perceptions of China and the Evolution of Russian-Chinese Relations Since the Eighteenth Century* (Armonk: M. E. Sharpe, 2003).

25. Today, 86 percent of foreign trade of the Far East district is undertaken with the countries of the Asia Pacific region (close to 30 percent with China, 30 percent with Japan, and 17 percent with South Korea) and is more than 32 percent of that of the Siberia district, while in the European regions, this share is only 7 to 17 percent. E. Saprykina, "The Russian Far East and Asia-Pacific Countries: Opportunities and Prospects of the Economic Integration," Vladivostok, Far Eastern National University, November 2008, <http://www.apru.org/awi/workshops/economic_integration/slides/Nov%208/1–2%20 Saprykina%20Elena%20FENU.pdf> (accessed February 10, 2010).

26. B. Lo, "A Fine Balance—The Strange Case of Sino-Russian Relations," *Russia-CEI Visions*, no. 1, April 2005.

27. M. Alekseev, "Migration, Hostility, and Ethnopolitical Mobilization: Russia's Anti-Chinese Legacies in Formation," in Blair Ruble and Dominique Arel (eds.), *Rebounding Identities* (Washington, DC: Woodrow Wilson Center Press, 2006).

28. B. Lo, "China and Russia. Common Interests, Contrasting Perceptions," *CLSA Asian Geopolitics* (London: Chatham House, May 2006).

29. R. Azizian, *Russia-India Relations: Stability amidst Strategic Uncertainty* (Asia-Pacific Center for Security Studies, October 2004).

30. N. Ul Haq (ed.), "Pakistan-Russia Relations," *IPRI Pakistan Papers*, June 30, 2007; H. Masood, "Pakistan-Russian Federation Economic Relations," *Journal of Political Studies*, Lahore, no date, no issue, pp. 35–50.

31. M. A. Smith, *Russia's Relations with India and Pakistan* (Defence Academy of the United Kingdom Conflict, Studies Research Centre, Russian Series, no. 4/24, August 2004).

32. Ibid., p. 4.

33. J. Cartwright, "India and Russia: Old Friends, New Friends," *CSIS South Asia Monitor*, no. 104, March 2007.

34. K. Subrahmanyam, "Indo-Russian Relations," *Dainik Jagran*, December 13, 2009.

35. "Russia and India Ready to Trade," *World Nuclear News*, December 5, 2008 <http://www.world-nuclear-news.org/NP_Russia_and_India_to_trade_0512082.html> (accessed February 10, 2010).

36. P. Dikshit, "India and Russia. Revisiting the Defence Relations," *IPCS Special Report*, no. 52, March 2008.

37. I. Facon, "Russie, Inde, coopération militaro-technique," *FRS Recherche et documents*, no. 8/2008.

38. Ibid., p. 56.

39. F. ur-Rahman, "Pakistan's Evolving Relations with China, Russia, and Central Asia," in Iwashita Akihiro (ed.), *Eager Eyes Fixed on Eurasia*, vol. 1, *Russia and Its Neighbors in Crisis* (Hokkaido: Slavic Research Centre Hokkaido University Sapporo, 2007), pp. 211–229.

40. Interviews conducted at the main Chinese think tanks and research centers working on the former Soviet Union in Shanghai, Beijing, Lanzhou, and Xi'an, September–December 2008.

41. M. Laruelle, S. Peyrouse, *China as a Neighbor. Central Asian Perspectives and Strategies* (Washington, DC: The Central Asia-Caucasus Institute, 2009).

42. Ibid., pp. 18–62.

43. N. Joshi, "India-Russia Relations and the Strategic Environment in Eurasia," in Iwashita Akihiro (ed.), *Eager Eyes Fixed on Eurasia*, vol. 1, *Russia and Its Neighbors in Crisis* (Hokkaido: Slavic Research Centre Hokkaido University Sapporo, 2007), pp. 196–210.

44. S. Akbarzadeh, "India and Pakistan's Geostrategic Rivalry in Central Asia," *Contemporary South Asia*, vol. 12, no. 2, 2003, pp. 219–228.

45. M. M. Puri, "Central Asian Geopolitics: The Indian View," *Central Asian Survey*, vol. 16, no. 2, 1997, pp. 237–268.

46. See detailed descriptions of Indian-Central Asian cooperation in Sébastien Peyrouse's chapter in this volume.

47. R. Alieva, "Tajiks Play Down India Base Reports," *IWPR Reports*, August 7, 2007, <http://www.iwpr.net/?p=rca&s=f&o=337703&apc_state=henh> (accessed February 10, 2010).

48. "India to Deploy Mig-29 Fleet at Tajikistan Air Base," *India Defence*, April 20, 2006, <http://www.india-defence.com/reports-1786> (accessed February 10, 2010).

49. S. Blank, "Russian-Indian Row over Tajik Base Suggests Moscow Caught in Diplomatic Vicious Cycle," *Eurasianet*, January 11, 2008, <http://www.eurasianet.org/departments/insight/articles/eav011108f.shtml> (accessed February 10, 2010).

50. B. Lo, "China and Russia. Common Interests, Contrasting Perceptions," p. 4.

51. B. Lo, *Axis of Convenience. Moscow, Beijing, and the New Geopolitics*.

CHAPTER 3

Central Asia-China Relations and Their Relative Weight in Chinese Foreign Policy

JEAN-PIERRE CABESTAN[1]

Ever since they declared their independence from the Soviet Union in late 1991, the five new Central Asian nation-states have kept Chinese foreign policy decision-makers busy—probably busier than they initially imagined they would be, not only because of the fresh challenges that this new reality triggered but also because of the unanticipated opportunities it brought in. There is also little doubt that, as far as Central Asia is concerned, security has been Beijing's top priority: border security, impact of these independent states on the situation in Xinjiang, as well as in Pakistan and Western Asia, in particular Afghanistan. The key role played by China in the creation of the Shanghai Group in 1996 and of the Shanghai Cooperation Organisation (SCO) in June 2001 has been a much publicized illustration of this persistent preoccupation. The 9/11 attacks and the United States' "global war against terrorism" have brought security issues into sharp focus in China's relations with Central Asian countries. However, the world's *sacred union* against Al Qaeda has also put the SCO under stress, compelling China to work out a more complex balance to the benefit of the latter, between, on the one hand, this multilateral arrangement and, on the other, the bilateral relations it had developed with each Central Asian country as well as with Russia. The growing danger of Islamic terrorism has also convinced Beijing to "walk on two legs" and place a bigger emphasis on economic cooperation and trade with Central Asia, a major factor contributing, in its view, to security on its western borders as well as in Xinjiang.

Having said that, China is pursuing a larger array of security objectives. Among them energy security has acquired unprecedented importance. Are China's energy security objectives in Central Asia competing or even conflicting with its more traditional security goals? What impact has China's increasing economic and trade influence in this region had on its security interests? Has it enhanced its security cooperation with Central Asia or has it, on the contrary, weakened it? How has this cooperation influenced and been influenced by the China-Russia "strategic partnership"?

This chapter will consider these questions and also try to reflect on the Xinjiang factor in China's relations with Central Asian countries. To what extent can this factor, perceived as a liability by most observers, be neutralized or even turned into an asset of China's policy toward Central Asia? Finally, the attempt here is to see China's relations with Central Asia in the larger context of its foreign policy and global international strategy, the hypothesis being that Central Asia has become a substantial feature of China's foreign policy but cannot pretend to fundamentally alter the country's well-established hierarchy of partnerships and priorities.

Security through Normalization of Relations

China was one of the first countries to recognize and establish diplomatic relations with the five new nation-states of Central Asia. One reason for the move, which tends to be forgotten today, was Beijing's fear that Taipei could take advantage of these freshly independent states to increase its number of diplomatic allies and enhance its international profile. Between 1992 and 1994, Taiwan, under its official name, the Republic of China, was able to operate a full-fledged consulate in Riga, Latvia. And in that same period of time, Taiwan itself, which had become newly democratic and was still strongly anti-communist under president Lee Teng-hui, began to rapidly develop affinities and multiply semi-official exchanges with several ex-members of the Soviet Bloc (Czech Republic, Poland, Ukraine), to the point that some of them even fancied forging official links with Taipei.[2]

However, Beijing's main concerns were elsewhere: all of a sudden, China realized that it would have to deal with three additional neighbors, Kazakhstan, Kyrgyzstan, and Tajikistan, controlling the long border and difficult terrain all the way from the Altai Mountains near Outer Mongolia in the north to Afghanistan's Wakhan corridor in the south. The Chinese government's most urgent priority was to convince its three new neighbors to carry on the border negotiations initiated with the Soviet Union under President Mikhail Gorbachev. Since none of them was ready to question the protocol of Tarbagatai (1868) or the treaty of Saint Petersburg (1881) in which the Russian and the Manchu empires agreed upon the delineation of their western border, negotiations resumed on the basis of these old accords. In October 1992, Beijing signed with Almaty (Kazakhstan's capital until 1997), Bishkek, and Dushanbe a first document approving "in principle" the borderline.[3] Of course, this agreement could not bring a final solution to all the pending disputes inherited from the past. For instance, it was only in 1998 that China and Kazakhstan signed a definite treaty on their 1,700-kilometer-long border, allowing the latter to keep 57 percent of the 944 disputed square kilometers (the borderline was changed only by 187 square kilometers).[4] Nevertheless, very early on, Beijing had made sure that its new neighbors recognized that Xinjiang was part of China and, as a consequence, could not in any way whatsoever support any movement aimed at recreating the short-lived Eastern Turkestan Republic.[5]

Neutralizing the impact of Central Asian states' independence on Xinjiang, however, proved to be a much more arduous task. In the 1980s, taking advantage of Deng Xiaoping's reforms and relative political opening as well as Afghanistan's fall into anarchy and gradual "Talibanization" after the Soviet withdrawal in 1988, pro-independence and pro-autonomy Uyghur organizations, although still mostly secular and inspired by Pan-Turkism, had become more active in Xinjiang.

Though China's People's Liberation Army (PLA) has always been able to crush any Uyghur military or peaceful rebellion, the establishment of three new and rather weak states (in any case much weaker than the defunct USSR) posed a new challenge to China. Of course, it should be reiterated here that none of the five Central Asian republics was either seeking or ready for independence. Independence was imposed upon them by the collapse of the Soviet Union and they were not really in a mood to propagate to Xinjiang and the Uyghurs their ill-prepared independent statehood. On the contrary, they were as keen as China to normalize relations, stabilize their borders, and develop economic cooperation.

However, could these newly formed governments, now dominated by the local Turkic elites (except in Tajikistan whose official language is identical to Persian), have control over Xinjiang's Uyghur activists who had taken refuge on the other side of the border? Moreover, China's biggest neighbor, Kazakhstan, includes a large local Uyghur community (estimated at between 300,000 and 500,000), which enjoys a genuine political and cultural autonomy (education system, universities, and newspapers). This community could shelter Xinjiang's pro-independence activists.[6] Kyrgyzstan also has a substantial Uyghur community that, although smaller (officially 1 percent of the population or around 50,000 but probably more), can more easily cross an ill-controlled border (at least on the Kyrgyz side).

It took some time for the Chinese government to convince its Central Asian counterparts to help; their assistance has on the whole remained limited and conditional. Linking this issue to the final conclusion of border treaties and the development of border trade and investments, the former was able to exert strong pressure on the latter. For instance, when the border treaty was signed in 1998, Almaty committed itself not to shelter Uyghur antigovernment "separatists" or pro-independence activists. In order to ease the acceptance of this external pressure, a year later, China signed with Kazakhstan a US$9.5 billion investment package that (already) included the construction of a 3,000-kilometer-long pipeline to Xinjiang.

As a show of goodwill, in the late 1990s, the Kazakh and the Kyrgyz governments banned all Uyghur political activities and organizations. Since then, only Uyghur "cultural associations" have been allowed to operate. Teaching on Xinjiang in Kazakh universities is today tightly controlled, restricted to the linguistic and cultural aspects of Uyghur civilization.[7]

Nevertheless, these bilateral pressures were not perceived by China as sufficient. In March 1996, the Chinese authorities started to rescind the slightly more tolerant policy toward the non-Han communities of Xinjiang that had been adopted after the beginning of the Dengist reforms in 1979 and embarked on an "assimilationist" policy, imposing much tighter control on Uyghur cultural and religious activities. Since then, Chinese authorities have curtailed and strictly managed pilgrimages to Mecca as well as enhanced the Uyghurs' "sinicization" and use of Mandarin Chinese, nurturing growing resentment among them. Moreover, Beijing has decided to openly encourage Han Chinese migrations, jeopardizing the Uyghur community's demographic domination of the Autonomous Region (9.5 million in 2008 or 45 percent of Xinjiang's population, as opposed to 9.2 million Hans or 43 percent, if migrant workers are included).[8] At the same time, in April 1996, China launched an unprecedented initiative and suggested to its three new neighbors as well as Russia the establishment of the Shanghai Group or Shanghai Five that, five years later, would be elevated to a proper multilateral security structure, the Shanghai Cooperation Organization (SCO).

Security through China-Russia-Led Multilateralism

Much has been written on the Shanghai Five and the SCO and other chapters in this volume have thoroughly dealt with this organization.[9] The focus here is on the main factors that led China to launch the "Shanghai Pact," as former Chinese president Jiang Zemin liked to qualify it, and how the SCO overcame the 9/11 challenge.

China's New Multilateralism and Good Neighborhood Diplomacy

First, it must be stressed that the Shanghai Group was a Chinese initiative; for the first time, in April 1996, in the wake of the Taiwan Strait missile crisis and a period of high tension with Washington, Beijing went out of its way to propose the establishment of a multilateral group aimed at building military confidence, reducing forces, and improving border security with four other neighboring countries. This unprecedented initiative underscored a change of mind, a new approach to regional security on the part of the Chinese leaders and diplomats. Of course, growing tensions on the maritime front had traditionally led the Chinese government to actively work out peace and stability on the continental borders. However, its modus operandi had changed: in 1996, it not only promoted multi-polarity but also decided to embrace, at least on some issues and with a selected number of partners, multilateralism.

It could also be argued that, apart from China, the four other members of the Shanghai Group were all former republics of the ex-USSR, and Russia had kept close if not always cordial relations with the three Central Asian nations that joined this group (it even had some troops stationed in civil-war-torn Tajikistan)—hence the need to have the Russians on board.

But clearly, the creation of the Shanghai Five was also a quiet admission by China and its new neighbors that the challenges they were facing were too big to be dealt with bilaterally. In Afghanistan, the situation was rapidly deteriorating. The Taliban were strengthening their grip over a country (they would take control of Kabul in September 1996) that had become a magnet for all kinds of terrorist organizations (Al Qaeda was to move there in 1998), a rear-base for Xinjiang's Uyghur activists and a platform for narcotics production and exports. As early as in July 1998, in Almaty, the Shanghai Group decided to give priority to the common fight against what since then has been called in China's official press the "three evils": "separatism, extremism, and terrorism."

The other major factor that made the Shanghai Group possible was a rapid and rather unexpected improvement in Sino-Russian relations after the end of the cold war and the emergence of the "Unipolar moment" dominated by the United States. After the collapse of the Soviet Union and Russia's democratization, there was a period of uncertainty, in particular for China. However, it rapidly became clear that it was in Russia's interest not only to carry on the process of "renormalization" with China initiated by Gorbachev but also to push it further in order to attempt balancing what both countries perceived as the unchallenged U.S. domination of the Eurasian continent and "NATO expansion to the east." It was no coincidence that in April 1996, Jiang Zemin and Boris Yeltsin established in Beijing a "strategic partnership" between their two countries—again a Chinese concept encapsulating non-confrontational and cooperative relations—and later created in Shanghai the Shanghai Group.

It is clear that Russia as well as China's three Central Asian neighbors were enthusiastic about the Shanghai Group—Russia, because it served its security interests at home (Chechnya) as well as in Western Asia and contributed to consolidating its links with some of its former dependencies (although it could achieve this goal through other organizations, as indicated below);

Kazakhstan, Kyrgyzstan, and Tajikistan, because they saw in China a new partner able to balance Russia's domineering influence and provide them more economic benefits than their former master.

Five years later, two more similar steps were taken. In June 2001, at China's initiative, the Shanghai Group turned into a proper organization and welcomed Uzbekistan, the other large country in Central Asia (population 27 million), sharing what was then a poorly controlled border with Afghanistan. Far from an integrated alliance, the SCO immediately became for Beijing not only a model of the flexible type but also a compelling and comprehensive multilateral regional security arrangement that the post–cold war world needed.[10] It included a strong security dimension, allowing the SCO members' law enforcement agencies to better interact and cooperate in their fight against terrorism, arms smuggling, drug trafficking, and illegal immigration. However, it also encompassed foreign policy, economic factors, and cultural facets that would gradually be strengthened over the years, at the instigation again of China.

A month later, on July 16, 2001, China and Russia signed an important "treaty of good neighbourliness and friendly cooperation" (*Zhong'E mulin youhao hezuo tiaoyue*). The treaty buried once and for all the long-denounced and abolished 1950 friendship treaty signed by Mao Zedong and Josef Stalin and heralded a new type of close but flexible partnership that would later be applied to many other countries or groups of countries, including the SCO itself. In 2007, the SCO concluded a "treaty of long-term good neighbourliness and friendly cooperation" (*changqi mulin youhao hezuo tiaoyue*) drafted on the Sino-Russian model.

The SCO and the Sino-Russian treaty were also part of a new and more general strategy that would rapidly become a prominent Chinese foreign policy objective: good neighborhood or "periphery diplomacy" (*zhoubian waijiao*). It was aimed at stabilizing and improving relations with all of China's neighbors, including the most contentious ones such as Japan and India. This strategy restored, if not an ancient and traditional modus operandi—today's relations between China and its neighbors are much less symbolically unequal but much denser than in the past when some of the latter were the former's tributary states—at least part of the traditional and somewhat patronizing imperial discourse: "*yu lin wei shan, yi lin wei ban*" (be benevolent toward neighbors, turn neighbors into partners) or, as Premier Wen Jiabao put it in October 2003 when attending an ASEAN summit in Bali, "*mulin, anlin, fulin*" (good, peaceful, and prosperous relations with neighbors).[11]

How the SCO Overcame the Challenges of September 11

September 11 took the SCO by surprise and, it should be recalled, raised questions about its raison d'être and survivability.[12] All of a sudden, the

United States stepped into Central Asia and, with the open support of Russian leader Vladimir Putin, established military bases in Uzbekistan (Karshi-Khanabad or K2) and Kyrgyzstan (Manas) to organize its offensive against the Taliban and to topple them from power in Afghanistan.

Nevertheless, the SCO managed to adapt to the new environment and, in an ironic twist of events, embarked on a policy of cooperation with the United States and the West against terrorism. Taking advantage of the global war against terrorism, Beijing decided to link all of the East Turkestan separatist organizations, though they were mostly inspired by secular or moderate Muslim ideologies, to international Islamic terrorism, particularly Al Qaeda, artificially "unifying" them into a single violent terrorist movement.[13] In 2002, the Chinese government succeeded in persuading the Bush Administration to help add the obscure and minuscule East Turkestan Independence Movement (ETIM) to the UN list of terrorist organizations. More importantly, the SCO strengthened its own coordination work. In 2002, it adopted a common approach to the prevention of international terrorism and decided to hold joint military maneuvers.

Then, for the first time under the SCO auspices, a joint military exercise involving 300 PLA and Kyrgyz soldiers and border guards took place in Kyrgyzstan. A year later, a larger military exercise by soldiers of all SCO members, except Uzbekistan, was organized in both Kazakhstan and China; its objective was to better evaluate the SCO members' capacity to fight together against terrorism. The first military exercise in which all SCO members participated took place in Xinjiang and Russia (Cheliabinsk, north of Kazakhstan) in August 2007.

Simultaneously, the SCO organizational structure was enhanced. At the June 2003 SCO summit in Moscow, it was decided to create a secretariat (which opened in Beijing in March 2004) and an anti-terrorism intelligence center (which was to have been established in Bishkek in January 2004 but eventually opened in Tashkent a few months later). In 2004, the SCO opened itself to its regional neighbors and created an observer status; Mongolia immediately acceded to the organization, to be joined, a year later, by India, Iran, and Pakistan. And in November 2005, because of deterioration in the Afghan situation and its negative regional consequences, an SCO-Afghanistan Contact Group was set up.

The U.S. invasion of Iraq (2003), the "color revolutions" in Georgia (2003) and Ukraine (2004), and later the "Tulip Revolution" in Kyrgyzstan (Spring 2005) as well as the repression of the Andijan riots in Uzbekistan (May 2005), all alarmed both Russia and China and convinced them to partly modify the objectives of the SCO: at the Astana summit (July 2005), "non-interference in the domestic affairs of sovereign countries" became a top priority; coordination against what was now qualified by the whole SCO as the "three evils" (separatism, extremism, and terrorism) was also strengthened; and in a transparent call for the closure of American military bases in Central Asia, it was reasserted that regional security could be ensured by local actors without external intervention. In August, for the first time, the Russian armed forces and the PLA organized in Shandong (China), under the SCO umbrella, a large-scale joint military exercise whose goal was to isolate fighters in a fictional country torn by ethnic conflicts, pursuant to a supposed UN mandate.[14] And in November, both Moscow and Beijing applauded Tashkent's decision to close the American K2 base.

Despite all these achievements, for China, probably more than for other members of the SCO, long-term security could not be achieved through military and intelligence means alone but needed to be based on economic development and prosperity as well.

Security through Economic Cooperation

Extending a strategy it applied in Xinjiang from the early 1980s, admittedly with some success, Beijing has constantly included an economic dimension to its policy toward its western neighbors, since even before the creation of the SCO and the Shanghai Group. For instance, as early as in 1991–1992, the railroad connecting Urumqi to Alma Ata/Almaty—and then on to the rest of Central Asia, Russia, and Europe—was opened to freight and then passengers. It completed an old project interrupted in 1959 by the Sino-Soviet rift. And after Central Asian states gained independence, the Chinese government more actively pursued trade and economic cooperation with those neighbors, gradually opening an increasing number of border posts and encouraging China's state-owned companies to invest in infrastructures and later in energy in the region. This new strategy has also accelerated Xinjiang's metamorphosis from a cul-de-sac into a dynamic regional trade hub as well as an intermediary between China and Central Asia.

After having become an oil importer in 1993, China tried hard, for strategic reasons, to diversify its sources of energy products and limit its dependence not only on the unstable Middle East but also on imports through seas and straits still mostly controlled by the U.S. Navy. As a consequence, buying more oil and gas from Russia and other continental suppliers, such as Central Asian states, became a key security objective.[15]

After the SCO's establishment and the 9/11 attacks, this emphasis on economic and energy cooperation intensified. In 2003, Wen Jiabao went further and suggested turning the SCO into a "free-trade zone."[16] Since 2001, border trade between China and Central Asia, mainly through Kazakhstan and Kyrgyzstan, has increased rapidly, making the region's markets more and more dependent on Chinese products.[17] Trade has also contributed to intensifying people-to-people contacts and exchanges between the Chinese (and especially Xinjiang) and

Central Asian societies. Meanwhile, Chinese investments in Central Asia have grown steadily, particularly in the energy sector and mainly in Kazakhstan.

Today, Kazakhstan still accounts for around 80 percent of Chinese commercial relations with, and probably also investments in, Central Asia (in 2008, Kazak trade with China was worth US$15 billion). In 2006, Chinese oil companies (CNPC, Sinopec, CNOOC) with an output of 18 billion tons controlled 26 percent of Kazakhstan's oil production.[18] In the energy sector, China (in particular CNPC) has also developed more recently an active cooperation with Turkmenistan, Uzbekistan, and Kazakhstan that has included the construction of a 1,800-kilometer-long gas pipeline. In July 2009, it was announced that China would grant Turkmenistan a US$3 billion loan to develop its South Yolotan gas field.[19] Also, in 2010–2011, the Atyrau-Alashankou 3,000-kilometer-long pipeline linking Kazakhstan's Caspian Sea fields (Kashagan and Tengiz in particular) to the Xinjiang border should be completed, boosting China's oil imports from Central Asia (4.25 million tons in 2007—2.8 percent of Chinese imports and 7 percent of Kazakh exports).[20]

Strategic partnership with Russia, the establishment of the SCO, and good neighborly policies as well as economically profitable relations not only with Kazakhstan, Kyrgyzstan, and Tajikistan, but also with other Central Asian nations, including Turkmenistan, have clearly contributed to enhancing China's both security and diplomatic influence beyond its western borders. However, China's growing economic and political influence in Central Asia has also led to fresh challenges in this part of the world and in its relations with Russia. These difficulties have forced Beijing to continue to give priority to its bilateral ties with its neighbors to the detriment of the SCO. Successes in Central Asia have not really helped Beijing to alleviate its problems with Xinjiang's Uyghur community. Conversely, these problems have, if not marred, at least complicated China's relations with this region and the Muslim world. In other words, open to all sorts of competing influences, Central Asia is far from becoming a new tributary of the world's second largest economy and power.

The SCO's Weaknesses

The Chinese government may have expected too much from the SCO, and some analysts even more. Stated differently, China needs to rely mainly on its bilateral relations with each Central Asian state rather than the SCO to pursue its objectives, defend its interests, or simply get things done. As indicated earlier, the SCO's good health depends greatly on the quality of the Sino-Russian relationship. This relationship has remained a complex one, filled with both cooperation and competition, (partly) common political values and jealousy, as well as shared interests and suspicion.[21]

In Central Asia, Moscow sees Beijing as a powerful and ambitious competitor, armed with the potential to end its traditional domination. As a consequence,[22] Russia considers the SCO as a marginal complement to the security or economic multilateral organizations in which most or all Central Asian nations participate, namely the Commonwealth of Independent States (CIS), the Collective Security Treaty Organisation (CSTO), and the Eurasian Economic Community (EurAsEC). The SCO Anti-Terrorism Centre established in Tashkent in 2004 does not share much intelligence with the Bishkek-based CIS Anti-Terrorism Centre managed by the Russians and to which five of the six SCO members (except China) belong. And it remains uncertain whether the closer coordination between the SCO and the CSTO initiated in 2007 will yield concrete results in terms of intelligence sharing, not only because of Russia but also owing to Central Asian countries' lack of enthusiasm to fully cooperate among themselves on these sensitive matters. Besides, Russia has refused to co-finance with China an "anti-financial crisis" stabilization fund. Later, China unilaterally enhanced its own economic influence in Central Asia (creating a US$10 billion fund), with Russia continuing to provide its dwindling assistance through the EurAsEC.[23]

The small scale of the military exercises conducted under the auspices of the SCO and the little practical benefit from them offer another illustration of the low degree of trust among its members and the organization's weak integration. True, the 2007 joint maneuvers held in Southern Siberia involved more troops (6,500 soldiers, including 2,000 Russian and 1,700 Chinese) than former exercises. However, in an obvious sign of distrust, the PLA tanks were not allowed to transit through Kazakh territory and had to be sent by train all the way through Manchuria and Eastern Siberia. And to date, the SCO has been unable to launch any military operation against Uyghur "terrorists" or "separatists" or even drug dealers outside Xinjiang borders, let alone within the region.[24] Moreover, there are numerous other disagreements within the SCO, not only among the Central Asian states, but also between Russia and China. The June 2009 border incident between Kyrgyzstan and Uzbekistan in the Ferghana Valley, probably involving some Al Qaeda–related movements, is one of the most recent examples of these frictions. And after the August 2008 Georgian crisis, Moscow was unable to persuade the other SCO members not only to recognize the two new states it has carved out from Georgia (Abkhazia and South Ossetia) but also to clearly endorse its policy. Chinese and Central Asian states' obsession with territorial integrity prevented them from demonstrating any solidarity with Russia on this issue.

Some may argue that the SCO brings to its Central Asian members the kind of multilateral setting and quality web they are unable to weave by themselves. But again, more familiar to them and sharing the same language, the above-mentioned Russia-led multilateral organizations, or simply the

Russian government, still appeal to these nations as the best loci of communication among themselves when bilateral relations cannot deliver or are stuck in unsolvable intricacies.

This is not to conclude that the SCO is totally irrelevant. It is a useful forum that has contributed to propagating China's approach to multilateral and regional security and economic cooperation. It has also helped Beijing rein in the organization of "separatist" and "terrorist" activities at the doorstep of Xinjiang. Actually, it can be argued that China likes the SCO as it now is, rather flexible and limited to Central Asia and Russia. Although it is trying to persuade Turkmenistan to join the SCO, with no success thus far, it has remained unwilling to expand the organization to its observers because of the danger of drowning it in a deeper ocean of conflicting interests and agendas. Nevertheless, in spite of its regular summits, its multiple consensual statements and agreements, and its continued survival, the SCO is far from being the main driver of China's growing influence in Central Asia. Beijing's successes in this region have been based rather on a bilateral diplomatic and economic activism adapted to each of its partners. The SCO has hardly helped to keep Uyghur- and Xinjiang-related opposition activities (violent or peaceful) under control. In fact, many dimensions of the problem are beyond its control, as they relate to events in Afghanistan or Pakistan, and probably to a larger extent in Xinjiang itself, intensified by the Chinese Communist Party's own rule.

China Facing Central Asia's Diversity and the Xinjiang Issue's Internationalization

Central Asia is a region made up of very diverse entities whose borders have in most cases been artificially delineated by its former Russian and Soviet colonizers and occupiers.[25] Moreover, among these entities, there are well-known cultural and political differences that have determined for each of them highly distinctive sets of foreign relations. On the whole, China has well understood this complicated reality and used it to its advantage. Nevertheless, this complexity has often, although to varying extents, hindered Beijing's influence and Chinese companies' development in this region. More importantly perhaps, Central Asian states' dominant Muslim-Turkic identity and solidarity have continued to mar their relations with a regime that has shown little tolerance of Xinjiang Uyghurs' quest for political, religious, and cultural autonomy.[26]

In weaker states such as Kyrgyzstan or Tajikistan, China has penetrated more easily. Chinese products flood the markets, Chinese engineers build roads to and in these countries, and some Chinese companies have made headways (as in Kyrgyz gold mines). However, these countries do not have much to offer. And the central authorities' lack of control over their territory complicates their relations with China. For instance, although pressured by Beijing, the Kyrgyz government arrested in August 2009 Dilmurat Akbarov, the leader of the Ittipak Uyghur society and his deputy Jamaldin Nasyrov; there is some doubt as to its capacity to keep in check Uyghur refugees' activities on its soil.[27] In fact, Kyrgyzstan is hardly China's most reliable partner in the region, having extended, after lengthy negotiations, the contract of the Manas U.S. base. Hence Beijing's recent decision to broadcast in Russian through its CCTV station in this country.[28] China seems to be fighting an uphill battle since Kyrgyzstan is the most pro-Western, albeit not the most liberal, state in the region—Bishkek is the seat of the American University in Central Asia, which trains many students from the whole region, in particular from the most isolated nation, Turkmenistan.[29]

Beijing's relations with Astana are much stronger and more stable. However, the Kazakh authorities are also wary of China's growing activism and are keen, on the diplomatic and trade fronts, to play China off against Russia or other countries (the United States, Europe, Japan). For instance, Kazakhstan has capped Chinese oil companies' acquisitions or access to oil fields (for example, preventing in 2003 Chinese participation in the North Caspian Sea Project controlled by a consortium of Western "majors"). And the Kazakhstan society—made up of 30 percent ethnic Russian and 70 percent Turkic, mainly Kazakh, ethnic groups—has noticeable cumulative resentment against Chinese commercial aggressiveness as well as Beijing's policy in Xinjiang. The Kazakh government has tried as much as possible to douse and downplay this resentment but it cannot totally ignore it and has discreetly expressed it to the Chinese authorities, in particular after the Urumqi riots of July 5, 2009. Moreover, activists affiliated to Rebiya Kadeer and the Washington-based Uyghur World Congress (UWC) are tolerated in Kazakhstan as long as they do not organize violent operations in Xinjiang from Kazakh soil.[30]

Uzbekistan, which is more closed and more Islamic but also more challenged by terrorist groups crossing from or having links in Afghanistan, is probably China's most enthusiastic partner in Central Asia. Indeed, both countries have good reasons to cooperate against the "three evils." Tashkent has approved CNPC's project to build the Yolotan (Turkmenistan)-Khorgos gas pipeline through its territory. However, anti-China feelings are strong, in particular among the intelligentsia and the business community that perceives Chinese companies' investments as spelling the death knell of local industry.[31] And one-fourth of the 27 million Uzbeks see themselves as enjoying very close blood relations with Xinjiang's Uyghurs (although they cannot be assimilated with Uyghur people) and can in no way remain indifferent to what happens in China's western autonomous region.[32]

Finally, ties between Turkmenistan and China have rapidly improved in the last few years. Initiated by the "Turkmenbashi" Saparmurat Niyazov before his sudden death in December 2006, these ties have been strengthened by the current president, Gurbanguly Berdymukhammedov. Both countries have established in the energy sector a privileged partnership, triggering much jealousy among other gas-thirsty nations. CNPC is to date the sole foreign company to have been licensed for onshore exploration and production in Turkmenistan. But this underpopulated country (5 million inhabitants) has remained too self-ostracized and remote to really become the anchor of China's interests in the region.

All in all, the side-effects of China's fast economic penetration in Central Asia as well as this region's ethnic solidarity with Xinjiang's Uyghurs are the main obstacles to any close and trustworthy partnership between the two sides, be it within or outside of the SCO. In addition, the UWC's peaceful strategy and support for a meaningful political autonomy in Xinjiang, as opposed to full independence, can sit well with China's main Muslim partners such as Iran, Saudi Arabia, and Turkey. Although Central Asian nations, SCO members in particular, must show restraint and discretion as to their real inclinations, there is much evidence to conclude that in the long term they see the UWC's plans as the best way to stabilize Xinjiang's relations with both China proper and the Central Asian as well as Western Asian regions.

In addition, these complex relations and realities raise a number of questions regarding the security of China's continental energy supply line. Can the respective Central Asian countries concerned and the Chinese authorities really guarantee the security of the long pipelines? Although the Turkmen and Kazakh territories appear today to be the safest ones, what about those in Uzbekistan and Xinjiang? Closely connected to Al Qaeda and Afghanistan, Islamic extremist organizations in Uzbekistan, such as the Islamic Movement of Uzbekistan, a structure that also includes a number of Uyghur activists, may be tempted to target the China-built pipeline, particularly as a response to developments in Xinjiang they disapprove of.[33] And as the situation in Xinjiang is far from showing signs of improvement, some local Uyghur radical movements may harbor similar destructive thoughts. In fact, since the July 5, 2009 anti-Chinese riots in Urumqi, Al Qaeda has for the first time identified China as an enemy. Al Qaeda in the Islamic Maghreb has called for revenge and seems ready to hit China's interests in Algeria and elsewhere.[34] More recently, Al Qaeda leader Abu Yahia Al Libi, who is thought to be based in Afghanistan or Pakistan, has called on Uyghurs to prepare for a "holy war" against the Chinese government in Xinjiang.[35]

The PLA is obviously aware of this new menace; however, it remains to be seen whether it will be able to preempt these terrorist acts from taking place, inside or outside of Xinjiang, and especially to protect long pipelines in China's immense western region (of over 1.66 million square kilometers).[36]

Conclusion: Central Asia in China's foreign Policy

The collapse of the Soviet Union, Central Asian states' emergence as independent nations, and the establishment of the "Shanghai Pact" have all noticeably rebalanced China's foreign policy. Xinjiang's isolation is over and China has relinked with its western neighbors and, for the first time since the nineteenth century and arguably since the battle of Talas in 751, is in a position to exert a genuine influence beyond the borders of its Turkestan protectorate. In other words, Central Asian countries have become for China more substantial and reliable partners.

However, Central Asia's relations with China will remain affected by the situation in Xinjiang as well as in Western Asia, in particular Afghanistan and Pakistan, preventing a full and trustworthy partnership from taking shape. And competition with Russia and others in Central Asia will certainly go on. As a matter of fact, this region has remained to a large extent a steppe open to the four winds and Beijing cannot take for granted that the successes it has achieved there will last into the long term. Not only Russia, but also the United States, the European Union, Japan, and South Korea (which enjoys the benefit of being able to capitalize on the Korean diaspora deported there by Stalin before World War II) are active in Central Asia and the local governments have been on the whole keen to take advantage of this set of pluralist influences and interests.[37] Moreover, the lingering ambiguities of Central Asian governments' and societies' attitude toward China and its atheist rulers will continue to complicate their bilateral relationship with Beijing as well as hinder the role of the SCO.

More importantly, China's foreign policy priorities have been and will probably continue to be elsewhere, on its maritime borders and beyond: the United States, Japan, the Koreas, Taiwan, and South East Asia before Russia, the European Union, India, Africa, and Latin America. China's trade with Central Asia has remained rather modest as least for the former; its imports of energy products, even when the two trans-regional pipelines are in full operation, will not be able to supply China more than 10 percent of its oil needs (as opposed to 30 percent of its gas, a much less crucial item that will probably continue to represent less than 5 percent of its overall energy consumption).

In this sense, in spite of the revival of China–Central Asia ties, the ancient and traditional strategic and security choices imposed upon the Chinese rulers cannot recover any relevance. Difficulties with and insecurity around its continental neighbors are the true irritants for Beijing. Nevertheless, the major forces able to balance or perhaps thwart China's rise are absent; they will continue to be concentrated on its maritime front, forcing the PLA to accelerate its modernization, especially its capacity to project forces in its surrounding seas as also across the oceans, in order to better secure its main sources of supplies or exports as well as the growing political and economic interests it needs to defend around the world.

Notes

1. Professor in Political Science, Head of Department, Hong Kong Baptist University.

2. C. Tubilewicz, *Taiwan and Post-Communist Europe. Shopping for allies* (London & New York: Routledge, 2007).

3. Two years later, in September 1994, Russia and China signed a border treaty delineating their 55-kilometer-long border between Kazakhstan and Mongolia.

4. <http://www.cacianalyst.org/?q=node/367> (accessed February 16, 2000).

5. With Kyrgyzstan, the negotiation took more time and a border treaty was signed only in 2002. In this accord, Kyrgyzstan agreed to transfer 900 square kilometers to China, sparking large protests. This is understood as one of the causes of the Tulip Revolution and the ousting of President Askar Akayev in 2005. It was also in 2002 that the final delineation of the Sino-Tajik border, based on the status quo, was agreed upon when presidents Jiang Zemin and Emomali Rachmonov signed in May the Supplement to the Boundary Agreement between the two countries concluded in August 1999.

6. This community is also politically represented; the current Kazakh prime minister Karim Massimov is widely referred to as an ethnic Uyghur.

7. J. A. Millward, *Eurasian Crossroads. A History of Xinjiang* (New York: Columbia University Press, 2007), p. 337. Interviews, Almaty, June 2008 and 2009.

8. The Hans amount to 8.5 million or 40 percent of the Xinjiang population without the migrant workers, calculated from Millward's data, *Eurasian Crossroads. A History of Xinjiang*, pp. 341–348. According to Chinese official statistics, at the end of 2008, the Xinjiang population totaled 21.3 million; *Beijing Review*, August 26, 2009; according to China's 2005 census, it amounted to 20.3 million: The Uyghurs made up 45.9 percent (9.2 million), Hans 39.6 percent (8 million), Kazakhs 7.03 percent, Huis 4.44 percent, Kyrgizs 0.85 percent and Mongols 0.85 percent. <http://english.mofcom.gov.cn/aroundchina/xinjiang.shtml> (accessed December 28, 2009). Some experts claim that in Xinjiang the Hans have now overtaken the Uyghurs in terms of population numbers.

9. On the SCO, cf., for example, M. Lanseigne, *China and International Institutions. Alternate Paths to Global Power* (London & New York: Routledge, 2005), pp. 115–142.

10. All decisions within the SCO are reached by consensus. The SCO treaty in Chinese and Russian as well as other SCO related documents can be found on its website at <http://www.sectsco.org/CN/show.asp?id=99> (accessed February 12, 2010).

11. *Xinhua*, October 8, 2003. Cf. also *Huanqiu shibao* (Global Times), December 12, 2003.

12. M. Oresman, "The SCO: A New Hope or to the Graveyard of Acronyms?" *PacNet*, no. 21, May 22, 2003.

13. J. A. Millward, *Eurasian Crossroads. A History of Xinjiang*, pp. 339–340.

14. I. Facon, "Les relations stratégiques Chine-Russie en 2005: La réactivation d'une amitié pragmatique," *Notes de la FRS* (Paris: Fondation pour la Recherche Stratégique, January 20, 2006).

15. T. N. Marketos, *China's Energy Geopolitics. The Shanghai Cooperation Organization and Central Asia* (Abingdon & New York: Routledge, 2007), pp. 20–30; K. Wu, "China's Energy Interests and the Quest for Energy Security," in E. Van Wie Davis, R. Azizian (eds.), *Islam, Oil, and Geopolitics, Central Asia after September 11* (Lanham, MD: Rowman & Littlefield, 2007), pp. 123–143.

16. L. Rhett Miller, *New Rules to the Old Great Game: An Assessment of the Shanghai Cooperation Organization's Proposed Free Trade Zone*, Baltimore, School of Law, University of Maryland, *Maryland Series in Contemporary Asian Studies*, no. 3, 2003.

17. G. Raballand, A. Andrésy, "Why Should Trade between Central Asia and China Continue to Expand?" *Asia Europe Journal*, vol. 5, no. 2 (2007), pp. 235–252. B. Kaminski, G. Raballand, "Entrepôt for Chinese Consumer Goods in Central Asia: The Puzzle of Re-exports through Kyrgyz Bazaars," *Eurasia Geography and Economics*, vol. 50, no. 5 (2009), pp. 581–590.

18. S. Peyrouse, "Chinese Economic Presence in Kazakhstan: China's Resolve and Central Asia's Apprehension," *China Perspectives*, no. 3 (2008), pp. 34–49.

19. Turkmenistan produces 80 billion cubic meter of gas a year, cf. *AP*, June 6, 2009. Interviews, Ashgabat, June 2009.

20. S. Peyrouse, "Chinese Economic Presence in Kazakhstan."

21. Cf. J.-P. Cabestan, S. Colin, I. Facon, M. Meidan, *La Chine et la Russie. Entre convergences et méfiance* (Paris: Unicomm, 2008).

22. Here it may be noted that Russia considers it no longer in its interests to support Uyghur separatists as it was doing until the late 1970s when the Sino-Soviet rift ran deep.

23. A. Cooley, "Cooperation Gets Shanghaied," *Foreign Affairs*, December 14, 2009, <www.foreignaffairs.com/> (accessed March 2, 2010).

24. Many drug traffickers use the porous border between Afghanistan and Tajikistan as a route to Russia or China. The SCO has not even tried to secure that border.

25. O. Roy, *La nouvelle Asie centrale ou la fabrication des nations* (Paris: Seuil, 1997).

26. C. Zambelis, "Xinjiang's Crackdown and Changing Perceptions of China in the Islamic World," *China Brief*, vol. 9, no. 16, August 5, 2009, pp. 4–8.

27.. "Uyghur Diaspora Faces Government Pressure in Kyrgyzstan," *Eurasia Daily Monitor*, vol. 6, no. 156, August 13, 2009.

28. *Eurasia Daily Monitor*, vol. 6, no. 186, October 9, 2009.

29. At least until fresh restrictions were imposed by Ashgabat in July 2009; Interviews, Bishkek, June 2009.

30. Interviews, Almaty, June 2009. For instance, after the July riots in Urumqi, Kazakh activists were allowed to "demonstrate" in a theatre in Urumqi: Interview, October 2009.

31. A. Ilkhamov, "Profit, Not Patronage: Chinese Interests in Uzbekistan," *China Brief*, September 27, 2005, vol. 5, no. 20.

32. The Uzbek language is closer to Uyghur than any other Central Asian Turkic language; interviews, Almaty, June 2008 & June 2009.

33. J. A. Millward, *Eurasian Crossroads. A History of Xinjiang*, p. 338.

34. C. Zambelis, "Xinjiang's Crackdown and Changing Perceptions of China in the Islamic World," p. 8.

35. *CNN*, October 8, 2009.

36. P. J. Smith, "China's Economic and Political Rise: Implications for Global Terrorism and U.S.-China Cooperation," *Studies in Conflict and Terrorism*, vol. 32, no. 7 (2009), pp. 627–645.

37. E. Wishnick, *Russia, China and the United States in Central Asia: Prospect for Great Power Competition and Cooperation in the Shadow of the Georgian Crisis* (Carlisle, PA: Strategic Studies Institute, U.S. Army War College, 2009).

CHAPTER 4

An Elephant in a China Shop? India's Look North to Central Asia... Seeing Only China

EMILIAN KAVALSKI[1]

The evaluation of New Delhi's agency in Central Asia attests to the region-alization of India's post–cold war foreign policy—that is, the development of distinct external policies toward different global regions. In this respect, the discourses of India's relations with Central Asia offer insights into the coun-try's strategic culture and the modes of security governance that it fashions. It has to be noted from the outset that this chapter undertakes an assessment of the narrative construction of India's involvement in Central Asia. The claim is that by focusing on the discourses of foreign policy making, this mode of analysis offers the opportunity to simultaneously experience and deduce *the ingredients* that go into the articulation of New Delhi's external outlook.[2] In other words, the discursive engagement with India's interactions in Central Asia not only indicates that India has a "strategic will" but also probes "how [New Delhi] will manifest it in action."[3]

During a visit to Turkmenistan in September 1995, the then prime minis-ter, P.V. Narasimha Rao, announced that "for India," Central Asia is an area "of high priority, where we aim to stay engaged far into the future. We are an independent partner with no selfish motives. We only desire honest and open friendship and to promote stability and cooperation without causing harm to any third country."[4] Rao's proclamation offers a glimpse into the discursive genesis of the "Look North" policy—the narrative framework of India's rela-tions with Central Asia. As its appellation suggests, the Look North policy strives to emulate the logic and achievements of India's "Look East" approach to Southeast Asia, which appears to have indicated that India has a strategy whose "footprint reaches well beyond South Asia."[5] Thus, the discourses of the Look North policy come to demonstrate that India is able to "break out of the claustrophobic confines of South Asia."[6]

The chapter proceeds with a narrative assessment of the Look North policy. The investigation draws attention to the significance of the post–cold war trajectories of India's foreign policy making on its relations with Central Asia. The analysis indicates that for the better part of the 1990s, India's foreign policy formulation has remained in the grips of conceptual tensions, strategic uncertainty, and geopolitical constraints. This attitude appears to have altered as a result of the May 1998 nuclear tests. The detonations have provoked a discursive overhaul of India's international relations premised on a much more assertive (if not aggressive) foreign policy stance. This study also reveals that the 1998 nuclear tests have impacted the narrative construction of the Look North policy. With this setting as the background, the chapter details the idiosyncrasies of India's post-1998 relations with Central Asia. Such contextualization makes possible the engagement with the discursive modalities of India's encounter with China in Central Asia. Such an employment informs the complex construction of Beijing as a *partner* in, a *threat* to, and a *model* for the Look North policy. Reflecting on this experience, the chapter concludes by asserting the lack of *influence* in India's Central Asian agency. The contention is that New Delhi's international image has few appealing attributes that regional states might be tempted to emulate.

The Narrative Outlines of the Look North Policy

Most commentators maintain that India's engagement with Central Asia is a function of the country's historical interactions with the region. Thus, the "long-standing historical ties encompassing the political, cultural, economic, and religious dimensions" constitute the basis for the international relations between New Delhi and the individual Central Asian states.[7] Commentators have thereby intimated that "the return of Central Asia to the world" has tended to "generate considerable romance" in New Delhi about rekindling the historic links between India and Central Asia.[8]

These (re)visions of New Delhi's involvement in the region have been underpinned by the conviction that "the glorious history of Indian cultural and intellectual achievement"[9] offers a shortcut into the security governance of Central Asia. India's engagement in the region is, therefore, often framed in the language of a "civilizational state that should deploy its culture (part of its inherent greatness) as a resource or valuable diplomatic asset, and others [i.e., the Central Asian states] must become cognisant of the moral quality of the Indian foreign and strategic policy."[10] Such an attitude follows the cognitive map of a strategic culture, which assumes a position of superiority because of its "ageless and rich civilization." Rodney Jones argues that the collective consciousness of India's past bolsters its "natural claim to greatness." Hence, its prominence in international life is assumed to be an outcome of its legacy—an entitlement that does not have to be earned, proven, or demonstrated.[11]

In this context, Indian commentators assert that the articulations of the "Look-North" policy outline a "proactive and meaningful approach that accords top priority to Central Asia."[12] Thus, the narrative uses of the legacies of the past by Indian foreign policy elites disclose a strategy that aims "to remind the new generation in Central Asia that India is not new to them but rather a very old friend if seen in the historical perspective. It creates a "love at first sight" situation exhorting these past civilizational entities to recognize each other."[13] This conviction underpins the strategic rationale of New Delhi's relations with Central Asia. Thus, the Look North policy becomes an expression of India's aspiration "to promote a secular, multiethnic order in the region,"[14] by establishing itself as a model for the Central Asian countries.

It has to be remembered, however, that the narratives of the Look North policy did not emerge in a vacuum but were profoundly implicated in the post–cold war trajectories of India's foreign policy making. The formulation of a country's international interactions offers discursive platforms for the manifestation of national self-positioning on the world stage and the re-contextualization of historical narratives to the exigencies of the present. The following section offers a brief outline of the post–cold war trajectories of New Delhi's external affairs. This sketch then provides a context for the engagement with the post-1998 articulations of India's Look North policy.

The Post-1998 Assertiveness of India's Policymaking

The narratives of the Look North policy reflect the complex trajectories of India's post–cold war foreign policy. One needs to remember that although nearly universally perceived as an opportunity for promoting different visions of "new world orders," for India the crumbling of the Berlin Wall represented "the loss of an entire world."[15] As Abid Husain, the former Indian ambassador to the United States, perceptively remarked, "One economist described India as a tiger in a cage. When the cage is open, the tiger would show its real strength. The cage is now open but the tiger refuses to come out of the cage."[16] Thus, for the better part of the 1990s India's foreign policy making exhibits a palpable reluctance to venture out of the cognitive cages of the defunct cold war order.

The reorientation of New Delhi's external outlook had to confront several key predicaments. On a pragmatic/policy level, India had to formulate a new international strategy (i) in the absence of its erstwhile ally—the Soviet Union; and (ii) while acknowledging the failure of (Nehruvian) nonalignment. On a conceptual/strategic level, India's foreign policy making became frustrated by the increasing tension between "militarism" (i.e., coercive international stance) and "moralism" (i.e., cooperative international stance). Consequently, India's policymaking anxiety in the immediate post–cold war environment attests to the inability to meaningfully accommodate the desire for a more assertive role on the global stage while lacking the confidence that it *can* and *should* do so.[17] Not surprisingly, these pragmatic and conceptual dilemmas produced a visible "normative schizophrenia" in New Delhi's international outlook.[18]

It is this setting that reveals the centrality of the 1998 nuclear tests to India's post–cold war foreign policy making. To begin with, the testing of nuclear weapons is never a small affair. However, the actual event tends to grow in magnitude when the state undertaking the test has refused to formally accede to the nuclear nonproliferation regime and is in a nearly permanent state of attrition with one of its (equally armed) neighbors. The five nuclear devices detonated by India during May 11–13, 1998 at the Pokhran range confirmed this point. The tests set off widespread criticism and, at the same time, prompted concerns about a nuclear arms race in South Asia. India was seemingly unfazed by the international censure, and the brazen geopolitical discourse of India's nuclear tests indicated a marked departure from its previously noncommittal and largely conciliatory attitude. In a nutshell, *they were aimed to provoke.*

By flaunting its ability for "pre-emptive response,"[19] New Delhi has publicized its newfound foreign policy independence. This assertiveness seems to have been born out of a longstanding frustration with New Delhi's marginalization in the international system. In this respect, India's post-1998 foreign policy stance has revealed not only the mere operationalization of "improved tactics,"[20] but also a qualitatively different interpretation of both the country's role in international life and the character of the international system. Thus, the 1998 nuclear tests had to indicate the resolution of India's pragmatic/ policy and conceptual/strategic predicaments.

First, they emphasized the end of New Delhi's ambivalence between "militarism" and "moralism." The nuclear detonations reflected the strategic decision "to rely more on power politics and less on morality and unilateral restraint in the pursuit of Indian interests."[21] The contention is that the 1998 nuclear tests have projected in an "emphatic manner" to the rest of the world India's newfound "self confidence,"[22] "without being subservient to hegemonic powers."[23]

Second, the nuclear tests appear to have lifted the perception that India needs a great power ally to substitute the defunct Soviet Union. Instead, in the post–cold war world such alliances are considered constraining to the autonomy of India's international interactions—that is, an assertive (nuclear) foreign policy is also an *independent* foreign policy. It is in this respect that the 1998 nuclear tests have become a symbol for the tectonic shift in New Delhi's external outlook. In particular, they have revealed India's willingness to insert its national interest on the global agenda, regardless of the opinions and expectations of other international actors. In many respects, this trend has precipitated an "Indocentric" approach to foreign policy, centered on the belief that "'strength respects strength' in foreign policy."[24]

Third, the 1998 tests indicate a profound break with the peaceful international attitude of nonalignment. Instead, their nuclear language demonstrates a conviction that it is the willingness to use force that can ensure India's security objectives.[25] To many observers, India's swift response during the 1999 Kargil War with Pakistan seems to have vindicated such militarized foreign policy stance. Thus, a central feature of the post-1998 international outlook has been a criticism of the perceived "softness" of Nehru's "pseudo-secularism," which "twisted India's strategic culture into all kinds of absurdities" and ultimately led to "enfeebling a once fierce nation."[26] In this context, the post-1998 assertiveness of India's forward foreign policy approach intimates the conviction that national security can be achieved in "essentially unilateral terms."[27]

India's Relations with Central Asia after 1998

India's relations with Central Asia became one of the most conspicuous aspects of its foreign policy ambiguity during the 1990s. In this respect, the nuclear tests became a catalyst for fresh perspectives on New Delhi's agency in the region. The post-1998 foreign policy perception of Central Asia seems to be informed by the realization that despite proclamations of the region's "historical belonging" to India's "strategic neighbourhood," New Delhi has "not given sufficient attention to Central Asia"; consequently, "good intentions have not been converted into substantive relations."[28] India has, therefore, been relegated to "a mere spectator" of Central Asian politics.[29]

In an attempt to rectify such marginalization, "India's 'forward' Central Asian policy" in the post-1998 period has been construed "as an integral component of its growing military, nuclear, and economic power."[30] The perception is that New Delhi's relations with the Central Asian states reveal "the operation of a virtuous circle, with economic gains propelling an associated political push from India [which] produces both an improved country image and a country's standing in regional and multilateral fora."[31] Thus, it is the assertiveness of New Delhi's foreign policy stance that is expected to make India more appealing to Central Asian states.

In this setting, the stated overarching objective of India's Look North policy is the promotion of "peace and mutual prosperity."[32] This intent, however, has been buttressed by the twin ambitions of (i) maintaining "the democratic and secular ethos" of the region, because it "binds India and Central Asia together"[33] and (ii) evolving "measures that would safeguard the stability and integrity of Central Asian republics and save them from getting divided and opposing one another."[34] At the same time, confirming the pragmatism of its post-1998 foreign policy, India has also developed a policy resting on the principles of strategic bilateralism in an attempt to overcome its marginalization in the region. The following sections detail such narrative construction of a security governance mechanism for Central Asia embedded in the narratives of the Look North Policy.

Experience of Managing Diversity within a
Secular and Democratic Polity

Indian commentators have noted that the (violence accompanying the) dissolution of the Soviet Union and (especially) the Federal Republic of Yugoslavia has "eroded the legitimacy of multiethnic, multi-lingual, and multi-religious states."[35] This observation informs the (tacit) conviction that India is one of the few remaining countries that still maintains some of the characteristic features of the now defunct socialist federal arrangements.[36] Consequently, such a realization underpins the responsibility of its foreign policy making to assert the viability of India's state-building project by demonstrating relevance and experience in successfully managing its internal diversity through the institutional arrangements of a secular and democratic polity. In other words, India is not "multicultural by accident," but "multicultural by design."[37] Thus, India's strategic objective in the region is to "work for the rise and consolidation of democratic and secular polities in Central Asia, because the spill-over of the rise of religious extremism may threaten India's own internal stability and security."[38]

Yet, New Delhi's involvement in Central Asia is not about proselytizing the values of democracy (at least not explicitly), but about "letting regional states see what India stands for and what it offers."[39] In other words, India's involvement in Central Asia appears to be animated by a sense of international politics exercised through the "power of example"[40]—that is, although democracy is a desirable value, India (owing to its respect for the sovereignty of other countries) cannot integrate democracy-promotion into its external agency, because it is "a matter for other states to recognize [this] and act in accordance."[41]

Indian expectations, however, have been frustrated by the realization that, although "conversant in the art of governance,"[42] Central Asian states are experiencing a pronounced democratic deficit. Indian commentators list multiple (and, oftentimes, contradictory) rationalities in their explanation of the weaknesses of the region's democratic practices:

- First, there is a near-universal agreement that the Central Asian states "were ill-prepared for independence."[43] According to this assessment, the independence of Central Asian states was an outcome of "the exclusively subjective ambitions of a group of political leaders and their scramble for political power."[44]

- Second, despite their independence from the Soviet Union, the leadership of the Central Asian states is still "dominated by the former communist party ruling elites who have adopted democratic pretensions."[45] This makes it "difficult for [regional leaders] to appreciate something [i.e., democracy] that their cognitive process has never experienced."[46]
- Third, the nascent forms of authoritarianism in Central Asia have been contextualized within longstanding indigenous forms, institutions, and processes of power relations. In particular, the reference to the clan-mentality of Central Asian politics points to the quandaries of a political process in which the dynamics of "regional and tribal affiliations" trump the legitimacy of "political vision and program."[47]
- Fourth, the post-Soviet transition of the Central Asian states has been profoundly disrupted by the nexus between drug trafficking, organized crime, and Islamic fundamentalism.[48]

The confluence between these factors seems to infer that there is "little prospect for a radical departure from the present scenario."[49] Equally importantly, the *passivity* underpinning India's attempt to establish itself as a model for Central Asian statehood through the "power of example" demonstrates an unusual style of international interactions—one in which New Delhi does not attempt to establish relationships with other actors by deliberately engaging them in shared practices. Some have read this attitude as an indication of the absence of "a global objective" in Indian strategic culture.[50] At the same time, despite their underlying assertiveness, the discourses of the Look North policy are not completely free of the conceptual and policy inconsistency that marked India's pre-1998 engagement in the region.

Encouraging Regional Cooperation in Central Asia

Intertwined with the narrative modalities of secularism and democracy, the Look North policy also stresses the significance of regional cooperation to the stability and prosperity of Central Asia. The proposition of Indian commentators is that "India should try [to] forge a collective security arrangement and a collective project for the development of all the countries of the region regardless of their policy slant in favour of this or that great power."[51] New Delhi's call for "greater political cooperation among the Central Asian states" rests on the perception of a "host of common specificities" that they share.[52]

The insistence on maintaining the unity of Central Asian states reflects New Delhi's apprehension that without regional integration, history might be repeated and Central Asia may lose "its creative capacity [just like it did] during the sixteenth century, owing to its internecine warfare, internal instability, and external aggressive policy."[53] In this respect, there seems to be a significant level of disappointment among Indian commentators that "the political leadership of these countries has been unable to evolve a mind-set that could be truly characterized as Central Asian."[54] This failure tends to be explained through the pursuit of narrow personal gains by nepotistic state-elites, which (more often than not) are disguised under the narrative cloak of (ethno-) national interests. Thus, commentators have noted that the failure of Central Asian states to establish a robust framework for regional cooperation illustrates their "poor governance."[55]

The regionalization implicit in the discourses of the Look North policy exposes a conviction that "it is our [i.e., India's] purpose to engage more vigorously with an independent Central Asia through cultural structures."[56] Such contextualization indicates New Delhi's resentment of the regionalizing strategies of other international actors. In this respect, some Indian commentators have propounded that the alleged "homogeneity [of the region] is quite deceptive" and hinders the comprehension of the "diversity, which is articulated in many different ways" in the complex dynamics of Central Asian politics.[57] Thus, the suggestion is that India needs to accompany its regional approach with "country-specific" strategies targeting the individual Central Asian states.[58] This understanding informs the discussion of India's bilateralism in the region in the following section.

Making Use of Bilateralism in Central Asia

As already suggested, the narratives of the Look North policy indicate a desire to encourage the regional cooperation of the Central Asian states. Such proclamations notwithstanding, India's engagement in the region has been paralleled by a significant level of bilateral relations between New Delhi and the Central Asian capitals. Pragmatically speaking, the development of strong bilateral relations with some countries in the region reveals New Delhi's attempts to overcome the constraints imposed by its latecomer status in Central Asian affairs, which has further compounded the effects from the lack of a direct physical access to the region.[59] Thus, India has adopted an approach aimed at making up for lost time with respect to the Central Asian agency of other international actors (especially, as we will soon see, China).

In this respect, it is Tajikistan that—to all intents and purposes—has become the poster child of New Delhi's strategic bilateralism in Central Asia. The construction of Tajikistan as India's "gateway to Central Asia"[60] is of complex provenance in the narratives of the Look North policy. The hackneyed point of departure seems to be the observation of the "millennia-old" "civilizational relationship between Tajikistan and the Indian subcontinent."[61] At the same time, many commentators assert that Tajikistan is "the first Central Asian republic [to] realize the importance of building a broader national identity based on institutions" and not on an exclusive ethnicity—a state-building strategy allegedly akin to that of India.[62]

Strategically speaking, however, it is the shared perception of external threats that appears to motivate India's bilateral relations with Tajikistan. Indian commentators explain that the civil war that ravaged the country during the 1990s has been *"caused* by a skilful exploitation of the inter-regional/inter-clan rivalries by forces of Islamic fundamentalism supported by the Pakistan-backed Mujahideen in Afghanistan"—that is, it was "a spill-over of the victory of the Mujahideen armed groups in Afghanistan. The jobless Afghan *jehadis* found employment both in Tajikistan and in the Indian state of Jammu and Kashmir."[63] In this respect, from the point of view of New Delhi, India and Tajikistan having "suffered a lot on account of onslaughts of the dark forces of religious extremism active in Afghanistan *must work together.*"[64]

Thus, while demonstrating the balance-of-power attitude informing India's post-1998 policy toward Central Asia, commentators tend to establish the significance of New Delhi's logistic and military support for the anti-Taliban Northern Alliance through Tajikistan. Such assistance was articulated as a strategy for "strengthening Tajikistan's secular forces in their war against Islamic fundamentalism."[65] For instance, there have been allegations that India's military outposts in the country have been set up as early as the mid-1990s.[66] In 2000 India formally acknowledged its assistance in the establishment of a military hospital, framed as an offer of "humanitarian assistance," on the Tajik-Afghan border at Farkhor and the widening of a military airstrip near Dushanbe for transport aircraft.[67] More recently, India—still "quietly, very quietly"—has deployed at least one helicopter squadron at its "Ayuni" airbase in Tajikistan to bolster its already existing rapid-response capabilities.[68]

The discourses of the Look North policy legitimize this military outreach by maintaining that "the nation's strategic interests lie far beyond [its] borders"—a realization that is "compelling New Delhi to consider the possibility of sending troops abroad *outside of the UN framework.*"[69] Thus, India's military presence in Tajikistan becomes one of the clearest indications of the presumed assertive (nuclear) logic of its post-1998 foreign policy. In this respect, India's involvement in Central Asia exposes an underlying "revisionist" foreign policy stance—through which New Delhi aims to *revise* the existing patterns in its international environment in order to facilitate the exercise of its own agency.[70]

Thus, the intense ties with Tajikistan reveal India's attempt to carve out a space for its stakes in Central Asia. At the same time, such bilateral relations do not demonstrate a socializing propensity that might become the cornerstone of a more encompassing community of practice in the region. The suggestion, therefore, is that until India overcomes the constraints and tensions of its strategic culture, it is unlikely to develop a meaningful security governance mechanism that would sway Central Asia in tune with New Delhi's international attitudes (and desires). What has been particularly frustrating for the articulations of the Look North policy is that while India's longing for closer relations with Central Asia has largely remained unfulfilled, China in the meantime has managed to establish itself as an important partner for the region. The following section details the complex context of India's encounter with China in Central Asia.

In the Eye of the Dragon: India's Encounter with China in Central Asia

While "looking north" at Central Asia, India has recognized that it is not the only international actor looking for an opportunity to insert its agency into the region. In fact, it quickly became apparent that China is the other international actor that India needs to measure up to. The complicated history of their bilateral relations frames New Delhi's perceptions of Beijing's Central Asian agency. In particular, the 1962 Sino-Indian war continues to have a bearing on India's attitudes toward China. Observers have suggested that the memory of the war remains a visible scar of the profound "psychological trauma"[71] and "great shock"[72] from which India's foreign policy making might not have completely recovered. Thus, the encounter with China in Central Asia has stirred up the underlying "anxieties" of New Delhi's post-1998 assertive international stance: "How will China's newly-found power influence the way in which it perceives and deals with the outside world? What are its motivations, ambitions, and goals? How will China use its growing power and how will this impact on the interstate system, and, particularly, on the region."[73]

At the same time, there is a pervasive awareness in the narratives of the Look North policy that the multiplicity of contending dynamics of globalization has "shifted the global strategic landscape" in Asia into a "more complex strategic situation."[74] China's ability to establish the Shanghai Cooperation Organization (SCO) in 2001 has become one of the clearest indications for such alterations. Indian observers have been quick to recognize the uniqueness of SCO. For instance, Sujit Dutta has declared that it is "a huge departure for Chinese foreign policy. For the first time China has initiated a multilateral strategic partnership."[75] SCO has thereby not only enhanced the visibility of "China's economic and political interests in the region"[76] but also reaffirmed Beijing's "new self-image"—that is, it "no longer sees itself as the plaything of international politics, but [as an] autonomous pole of power and purpose."[77] SCO has, therefore, become the epitome of China's socializing propensity in world affairs. Thus Indian commentators have gradually come to recognize that SCO is emerging as *the principal basis for strategic interactions* between Central Asia and the big and medium powers that surround the region."[78]

It is this understanding that informs Indian interpretations of Beijing's agency in Central Asia. Such an encounter seems to confirm perceptions that India and China are gradually becoming the "fulcrum of Asia."[79] Yet, others have suggested that Beijing's Central Asian agency confirms that if Asia is ever to become a functioning region "it would be due to the role played by China... Thus, it would not be too much of an exaggeration to say that China defines Asia; there can be no Asia without China."[80] In this setting, India's encounter with China in Central Asia has instigated three distinct images of Beijing's regional agency: *partner*, *threat*, and *model*. Although not necessarily complementary, all three representations are integral to the narratives of the Look North policy.

China's Partnering in the Look North Policy

The suggestion of a possible partnership between India and China in Central Asia reflects the presumptions of their post-1998 bilateral relations. In this respect, as both countries "were becoming more confident of their identities and [foreign policy] independence [they] were [also] becoming far more balanced and forthcoming in appreciating each other's views and preferences."[81] Beijing's agency in Central Asia has thereby been interpreted as a *genuine attempt* "to allay the natural fears of its neighbors and to respond to their concerns."[82] At the same time, Indian commentators have appreciated that such an attempt was not driven by a desire to "control and influence" the region.[83]

Thus, while recognizing China's resource-driven agency in the region, the narratives of the Look North policy nevertheless acknowledge the benefits for Central Asian states from joining "the dynamic Asian heartland" and gaining access "to the markets of the Asia–Pacific economic ring."[84] It is in this context that some Indian commentators see the possibility of a nascent partnership between New Delhi and Beijing. It is argued that the Chinese "pipeline development fits very well as a viable energy supply route for Central Asian oil to India" through Xinjiang.[85] At the same time, Beijing's involvement in the region can assist not only with enhancing "the goodwill" between the government of India and China, but also with improving the "trans–frontier interaction between the people on both sides [which] would help in reviving the pre-1960s relations between the two countries."[86]

Another commonality between India and China emphasized by the discourses of the Look North policy is their shared aversion to the practices of external democracy-promotion advanced in Central Asia by various Western actors.[87] Thus, although the two countries do not have "identical views on human rights, [they] both agree that for developing countries, the most fundamental human rights remain the right to subsistence and the right to development. [New Delhi and Beijing] remain particularly opposed to the practice [of] using economic aid as an instrument for bringing pressure on certain countries."[88] In this respect, although India "urges its neighbors to practice democracy, [it] also wants that the people of each country decide on the system that suits them...It is the prospect of aggressive peddling of democracy in various parts of the world that makes India uncomfortable."[89]

Such a similarity of normative attitudes has urged a number of commentators to anticipate a pattern of partnership in the Central Asian interactions between India and China. In particular, the presence of shared appreciation of and subscription to the inviolability of the national sovereignty of states has been expected to bring their foreign policy perspectives closer to one another. For instance, "their attachment to traditional concepts of absolute sovereignty is strong, especially in matters pertaining to territory and security. Neither is amenable to external interference on military and arms-related matters; both oppose external inspections and interventions elsewhere, especially if undertaken without the benefit of UN validation."[90] However, the very dynamism that underscores this perception of commonality between New Delhi and Beijing also animates Indian threat-perceptions of China.

China as a Threat to the Look North Policy

Despite proclamations of friendship and pragmatic strategic partnership, the context of Central Asia indicates that both India and China find it difficult to exorcise the ghosts of the past from their international interactions. The narratives of the Look North policy suggest that the deeply engrained "apprehensions" between New Delhi and Beijing have made both of them very "cautious of taking [joint] initiatives."[91] India, in particular, remains wary of the "*uncertainty* [that hangs] over China's future political and military direction" and "how [China's] growing power [in Asia] would be used."[92] Thus, despite the altered international environment, many Indian observers see in China's relations with Central Asia merely a new version of its longstanding policy for containing India through the provision of assistance to states in New Delhi's strategic neighborhood.[93]

Such attitudes toward China's agency in Central Asia reveal that "Sino-Indian relations, despite fitful improvements, will remain competitive [because] the two states have divergent self-images and different political systems. They also each wish to emerge as major powers in Asia and beyond."[94] Said otherwise, "India's desire to play an independent role of its own...in the strategic balance of Asia is in direct conflict with China's ambition of making Asia *its area of influence*."[95] In this context, the feeling is that Beijing's "present posture [in Central Asia] cannot be viewed in benign terms—if China is being accommodating, it is only because the balance [of power] does not seem to be in its favor; once its capabilities grow, it may adopt more strident policies [in the region]."[96]

The perception of China as threat to the objectives of the Look North policy also interprets the development of SCO as part of Beijing's nascent desire for hegemony in Asia. This attitude reflects the context in which New Delhi was granted the status of observer country (rather than full membership) to SCO. It has been interpreted as a confirmation of Beijing's attempt to construct India as "China's junior partner" in the region—meaning that it "dances" (or at least is expected to "dance") to the tune of the "senior partner."[97] Thus, the realization that SCO has brought China and the countries of Central Asia "closer politically and economically"[98] has contributed to New Delhi's growing suspicion of Beijing's proclamations in favor of multilateralism.[99]

India has been particularly concerned by China's military assistance to Central Asian states—thus, the attitude of some observers is that in Central Asia "China now calls the shots and tries to [ride] roughshod in the economy on its own terms."[100] Taking these "misgivings" as a point of departure, Indian commentators have insisted that "Central Asians would be more comfortable" if New Delhi were to engage in a more proactive "balancing [of] China's increasing influence in the region."[101] The acuteness of India's threat-perceptions in the context of its encounter with China in Central Asia reflects the underpinnings of the assertive promotion of its post-1998 foreign policy.

China as a Model for the Look North Policy

The sections above have outlined the modalities of the Indian perception of China either as a prospective partner or as a looming threat to New Delhi's relations with Central Asian states. Such discursive formulations rest on the understanding that Beijing's regional agency is simultaneously "need-based" and "ambition-guided."[102] Reflecting on the shared ideational origins of these bifurcated foreign policy perspectives, the narratives of the Look North policy illustrate that these images of China are not contending, but concurrent—that is, India's encounter of China in Central Asia suggests that their bilateral relationship would involve "both competition and cooperation [as] real life is always more complex."[103]

Thus, by acknowledging the complexity of Sino-Indian relations, a number of commentators have suggested that New Delhi's encounter with Beijing's agency in Central Asia has produced the image of "*China as a role model*" for India's international agency.[104] The realization is that "China came from far behind and overtook India [and] its success offers useful lessons."[105] Thus, the narrative of the Look North policy indicates that if India is to become the great power that it proclaims to be, it needs to learn from (and perhaps even emulate) the model set-up by Beijing. This foreign policy perception reflects the suggestion that India's relations with Central Asia facilitate the emergence of discourses that engage in parallel assessment of New Delhi's agency vis-à-vis other international actors.

A number of these "lessons" relate to the structure, process, and content of India's external relations. For instance, Kishan Rana notes that Beijing's relations with Central Asia follow the "strategic and tactical suppleness" of its foreign policy pragmatism, which does not waste time on identifying "a threat or an opportunity" and instead unfolds far-reaching "strategic partnerships" (SCO, for instance). The coherence and cohesiveness of these initiatives rest on a Chinese "blend of soft power [that] is designed to produce an appealing image" for Beijing's agency and is, therefore, not a mere accompaniment to its "mainstream activities."[106] Thus, in contrast to India, China's initiatives in Central Asia indicate the development of a sophisticated "holistic view" of foreign policy making, a view that "embeds the state firmly within the interstate system as an organic and inseparable part, linking the fate even of the inside of the state to the fate or nature of its outside."[107]

The confrontation with China's "holistic" foreign policy in Central Asia has stimulated the recognition of the need to emulate Beijing's ability to "establish quickly an international reputation for being able to look after itself [and, thus] become a "great power," whereas India's potential remains unrealized."[108] This image of China as model rests on (i) "China's ability to take hard internal decisions as well as to face up to pressure from the West that has been lionized by sections within virtually the whole spectrum of public opinion in India"; and (ii) "China's emergence as an economic and military power of significance and manner in which it has been able to reform its economy without compromising on its security posture is also viewed with awe and admiration."[109] Thus, the encounter with Beijing's involvement in Central Asia has produced diverse assessments of its agency within the narratives of the Look North policy, all of which tend to reflect the difficulties in articulating a foreign policy strategy in a world marked by pervasive complexity.[110]

Conclusion: The Lack of Influence of the Look North Policy

The discussion of the narratives of the Look North policy confirms New Delhi's foreign policy desire that India becomes "*a kind of a model* for other countries."[111] The proclivity toward a narrative projection of India as a blueprint for Central Asian development has become a defining feature of the Look North policy. Yet, beyond its discursive framing, New Delhi appears to have been either unable or unwilling "to face up to the challenges and opportunities of a dynamic security environment."[112] As demonstrated, the confrontation with the reality of Central Asian interactions and the involvement of other international actors—especially, China—makes conspicuous that New Delhi has little (if any) influence in the region.

This chapter has, therefore, contended that the discursive proclamations of India's external agency in Central Asia have not been matched by comparable transformations in the ideational and institutional makeup of New Delhi's foreign policy formulation. Thus, despite the proliferation of discourses on India's rise to global prominence, the absence of a readily available Indian vision of global politics—a *Pax Indica*, if you will—prevents New Delhi from living up to the expectations generated by such narratives. Not surprisingly, therefore, India's perception of the strengthening of the Beijing-based SCO has further aggravated "New Delhi's discomfiture" with "China's growing acceptance as the next global power."[113]

The contention here is that such attitude has been spurred by the palpable hiatus between India's self-aggrandizing perception of its post-1998 external outlook and the less flattering evaluations (if not outright "pejorative stereotyping") by others.[114] More conspicuously, the absence of a meaningful power of attraction has prevented India's international engagement with Central Asia. Instead, India's security governance of the region has been "indistinct and incoherent"[115] because of its underlying "*lack of vision*."[116] This outcome has been underlined by "India's noticeable absence" from Central Asian politics.[117]

In other words, the proposition of this chapter is that India has a number of influential individuals and businesses, but it itself—as an international actor—does not have an *influential* foreign policy strategy that could establish it as an alternative to existing ones. Such an assertion should not be misinterpreted as an allegation that India does not have a respected and significant historical and cultural heritage. However, the international recognition of this legacy (just like the recognition of its nuclear capabilities) does not amount to *influence*—especially, influence that would be able to arouse a desire for emulation among other actors.

This qualification is necessary, because observers oftentimes tend to confuse the notions of "India as a rising power" and "Brand India"—that is, the latter is used to infer the former.[118] "Brand India" (as the term suggests) refers to the brands that export India's image abroad—from the cultural products of Bollywood to the entrepreneurial innovation of Bangalore. Too often, therefore, analysts take as their point of departure the assertion that India has "earned *international respect* and *influence*" through its booming economy, IT revolution, and nuclear weapons.[119] "Brand India" is taken not only to indicate the "awakening of a slumbering elephant,"[120] but also to demonstrate a unique "unshackling of the imagination"[121] that ushers in an "India unbound"[122] onto the global stage. The conceptual confusion between "India as a rising power" and "Brand India" underpins the tendency of Indian commentators to allege that their country has significantly impacted international trends, while external observers barely take notice of New Delhi's agency.[123]

The discussion of the narratives of the Look North policy has demonstrated that the discursive construction of India's current external affairs does not project a specific (if any) vision of world order that would distinguish it from the other participants in the "new great game." Consequently, the international identity of New Delhi has no distinct attributes that regional actors might be tempted to emulate. The implication then is not only that India might remain a "rising power" for longer than its pundits portend,[124] but also that the analysis of the cognitive framework of its strategic culture puts it in "the class of countries that are always emerging but never quite arriving."[125]

Thus, the question that transpires from the analysis of New Delhi's agency in Central Asia is whether India can "offer an alternative vision of a new world order"[126]—for instance, just like China appears to have done through the institutionalization of SCO. In other words, the analysis of India's external affairs still does not seem to offer a convincing response to the question of whether India "*can change enough*" to become a pole of attraction in an international environment marked by "extreme turbulence."[127] The current setting seems to suggest that India would retain its relative position of *no influence* for some time to come.

Notes

1. Lecturer in Politics and International Relations at the University of Western Sydney.
2. The methodology and inferences of this analysis follow the argument developed in E. Kavalski, *India and Central Asia: The Mythmaking and International Relations of a Rising Power* (London: I.B.Tauris, 2010).
3. S. N. Bal, *Central Asia: A Strategy for India's Look-North Policy* (New Delhi: Lancer Publishers, 2004), p. 35.
4. Quoted in S. D. Muni, "India and Central Asia: Towards a Cooperative Future," in N. Joshi (ed.), *Central Asia—the Great Game Replayed: An Indian Perspective* (New Delhi: New Century Publications, 2003), p. 110.
5. S. Saran, "Present Dimensions of the Indian Foreign Policy," in A. Sinha, M. Mohta (eds.), *Indian Foreign Policy: Challenges and Opportunities* (New Delhi: Academic Foundation, 2003), p. 115.
6. B. S. Gupta, "India in the 21st Century," *International Affairs*, vol. 73, no. 2 (1997), p. 309.
7. S. N. Bal, *Central Asia: A Strategy for India's Look-North Policy*, p. 4.
8. S. D. Muni, C. R. Mohan, "India's Options in a Changing Asia," in R. R. Sharma (ed.), *India and Emerging Asia* (New Delhi: Sage, 2007), p. 66.
9. U. Kachru, *Extreme Turbulence: India at the Crossroads* (New Delhi: HarperCollins, 2007), pp. 207–231.
10. S. K. Mitra, "The Reluctant Hegemon: India's Self-perception and the South Asian Strategic Environment," *Contemporary South Asia*, vol. 12, no. 3 (2003), p. 414.
11. R. W. Jones, *India's Strategic Culture* (Carlisle, PA: Defence Threat Reduction Agency, 2006), p. 7.
12. N. Joshi, "India's Policy toward Central Asia," *World Focus*, vol. 28, no. 335/336 (2007), p. 445.
13. B. M. Jain, *Global Power: India's Foreign Policy, 1947–2006* (Lanham, MD: Lexington Books, 2008), p. 210.
14. G. Dietl, "Quest for Influence in Central Asia: India and Pakistan," *International Studies*, vol. 34, no. 2 (1997), p. 142.
15. N. Menon, A. Nigam, *Power and Contestation: India since 1989* (London: Zed Books, 2007), p. 166.
16. A. Husain, "Opinion," *India Today*, March 15, 1993.

17. C. R. Mohan, *Crossing the Rubicon: The Shaping of India's New Foreign Policy* (London: Palgrave Macmillan, 2004), p. 208.

18. R. Kumar, "India as a Foreign Policy Actor: Normative Redux," in N. Tocci (ed.), *Who Is a Normative Foreign Policy Actor? The European Union and Its Global Partners* (Brussels: Centre for European Policy Studies, 2008), p. 212.

19. R. Basu, *Globalization and Indian Foreign Policy* (Jaipur: Vital Publications, 2007), pp. 207–222.

20. A. Narlikar, "Peculiar Chauvinism or Strategic Calculation? Explaining the Negotiating Strategy of a Rising India," *International Affairs*, vol. 82, no. 1 (2006), p. 74.

21. A. Kapur, *India—from Regional to World Power* (London: Routledge, 2006), p. 23.

22. *Hindustan Times*, "Sign of Self-confidence, Say Experts," May 12, 1998.

23. B. M. Jain, *Global Power: India's Foreign Policy, 1947–2006*, p. 220.

24. S. Chaulia, "BJP, India's Foreign Policy, and the 'Realist Alternative' to the Nehruvian Tradition," *International Politics*, vol. 39, no. 2 (2002), p. 221; C. R. Mohan, *Crossing the Rubicon: The Shaping of India's New Foreign Policy*, pp. 205–207.

25. K. Bajpai, "Indian Strategic Culture," in M. R. Chambers (ed.), *South Asia in 2020: Future Strategic Balances and Alliances* (Carlisle, PA: Strategic Studies Institute, 2005), p. 253.

26. S. Chaulia, "BJP, India's Foreign Policy, and the 'Realist Alternative' to the Nehruvian Tradition," p. 220.

27. A. Latham, "Constructing National Security: Culture and Identity in Indian Arms Control and Disarmament Practice," in *Contemporary Security Policy*, vol. 19, no. 1 (1998), p. 137.

28. J. N. Dixit, "Emerging International Security Environment: Indian Perceptions with Focus on South Asian and Central Asian Predicaments," in K. Santhanam, R. Dwivedi (eds.), *India and Central Asia: Advancing the Common Interest* (New Delhi: Anamaya Publishers, 2004), pp. 18–19.

29. S. Mahalingam, "India-Central Asia Energy Cooperation," in K. Santhanam and R. Dwivedi (eds.), *India and Central Asia: Advancing the Common Interest* (New Delhi: Anamaya Publishers, 2004), p. 133.

30. B. M. Jain, *Global Power: India's Foreign Policy, 1947–2006*, p. 210.

31. K. S. Rana, *Asian Diplomacy: The Foreign Ministries of China, India, Japan, Singapore, and Thailand* (New Delhi: Oxford University Press, 2007), pp. 62–63.

32. Y. Sinha, "India and Central Asia in the Emerging Security Environment," in K. Santhanam and R. Dwivedi (eds.), *India and Central Asia: Advancing the Common Interest* (New Delhi: Anamaya Publishers, 2004), p. 9.

33. N. Joshi, "Regional Economic Cooperation and Transport Links with Central Asia: An Indian Perspective," in K. Santhanam, R. Dwivedi (eds.), *India and Central Asia: Advancing the Common Interest* (New Delhi: Anamaya Publishers, 2004), p. 210.

34. S. K. Asopa, "India and Post-Soviet Asia: An Appraisal of India's Central Asian Policy," in J. N. Roy and B. B. Kumar (eds.), *India and Central Asia: Classical to Contemporary Periods* (New Delhi: Concept Publishing, 2007), p. 188.

35. J. N. Dixit, "Emerging International Security Environment: Indian Perceptions with Focus on South Asian and Central Asian Predicaments," p. 24.

36. E. Kavalski, M. Zolkos (eds.), *Defunct Federalisms: Critical Perspectives on Federal Failure* (Aldershot: Ashgate, 2008).

37. N. Manohoran, "Multiculuturalism in India: Diverse Dots in a Multiple Mosaic," in M. Mekenkam, P. van Tongeren, H. van de Veen (eds.), *Searching for Peace in Central and South Asia: An Overview of Conflict Prevention and Peacebuilding Activities* (Boulder, CO: Lynne Rienner, 2003), p. 330.

38. S. D. Muni, "India and Central Asia: Towards a Cooperative Future," pp. 98–99.

39. J. Jawahar, "India and Kazakhstan: Significant Changes," *SAPRA India*, February 24, 2002.

40. P. B. Mehta, "Still under Nehru's Shadow? The Absence of Foreign Policy Frameworks in India," *Indian Review*, vol. 8, no. 3 (2009), pp. 210–219.

41. R. W. Jones, *India's Strategic Culture*, p. 7.

42. S. K. Asopa, "India and Post-Soviet Asia: An Appraisal of India's Central Asian Policy," p. 173.

43. N. Joshi, "Issues in Central Asian Stability," in N. Joshi (ed.), *Central Asia—the Great Game Replayed: An Indian Perspective* (New Delhi: New Century Publications, 2003), p. 69.

44. S. N. Bal, *Central Asia: A Strategy for India's Look-North Policy*, pp. 301–302.

45. S. D. Muni, "India and Central Asia: Towards a Cooperative Future," p. 103.

46. Padma L. Dash, "Central Asia: Tulips Have Different Hews," in J. N. Roy, B. B. Kumar (eds.), *India and Central Asia: Classical to Contemporary Periods* (New Delhi: Concept Publishing, 2007), p. 190.

47. R. R. Sharma, "Issues Confronting Central Asia," *International Studies*, vol. 41, no. 4 (2004), p. 429.

48. C. Mahapatra, "The United States and the Asian powers," in R. R. Sharma (ed.), *India and Emerging Asia* (New Delhi: Sage, 2005), pp. 168–169.

49. Padma L. Dash, "Central Asia: Tulips Have Different Hews," p. 204.

50. P. B. Mehta, "Still under Nehru's Shadow? The Absence of Foreign Policy Frameworks in India."

51. S. K. Asopa, "India and Post-Soviet Asia: An Appraisal of India's Central Asian Policy," pp. 173–188.

52. R. G. Gidadhubli, "Politics of Oil and Natural Gas in Central Asia: Conflicts and Cooperation," in J. N. Roy, B. B. Kumar (eds.), *India and Central Asia: Classical to Contemporary Periods* (New Delhi: Concept Publishing, 2007), p. 171.

53. M. Haidar, "India and Central Asia: Linkages and Interactions," in N. Joshi (ed.), *Central Asia—the Great Game Replayed: An Indian Perspective* (New Delhi: New Century Publications, 2003), p. 260.

54. N. C. Behera, "Forging New Solidarities: Nonofficial Dialogues," in M. Mekenkam, P. van Tongeren, H. van de Veen (eds.), *Searching for Peace in Central and South Asia: An Overview of Conflict Prevention and Peacebuilding Activities* (Boulder, CO: Lynne Rienner, 2003), p. 210.

55. S. K. Pillai, "Border Conflicts and Regional Disputes," in M. Mekenkam, P. van Tongeren, H. van de Veen (eds.), *Searching for Peace in Central and South Asia: An Overview of Conflict Prevention and Peacebuilding Activities* (Boulder, CO: Lynne Rienner, 2003), p. 249.

56. M. K. Palat, "The 'Romance' of the Silk Road," *The Hindu*, January 9, 1999.

57. R. R. Sharma, "Political System and Democratic Discourse in Central Asia: A View from Outside," in J. N. Roy, B. B. Kumar (eds.), *India and Central Asia: Classical to Contemporary Period* (New Delhi: Concept Publishing, 2007), p. 123.

58. Padma L. Dash, "Central Asia: Tulips Have Different Hews," p. 205.

59. See E. Kavalski, "Partnership or Rivalry between the EU, China and India in Central Asia," *European Law Journal*, vol. 13, no. 6 (2007), pp. 839–856; E. Kavalski, "The Complexity of Global Security Governance," *Global Society*, vol. 22, no. 4 (2008), pp. 423–443.

60. S. K. Asopa, "Regional Interests in Tajikistan," in M. Singh (ed.), *India and Tajikistan: Revitalizing a Traditional Relationship* (New Delhi: Anamika Publishers, 2003), p. 153.

61. D. Kaushik, "A Decade of Independent Tajikistan: Some Reflections," in M. Singh (ed.), *India and Tajikistan: Revitalizing a Traditional Relationship* (New Delhi: Anamika Publishers, 2003), pp. 30–37.

62. R. R. Sharma, "State-building in Tajikistan: Problems and Prospect," in M. Singh (ed.), *India and Tajikistan: Revitalizing a Traditional Relationship* (New Delhi: Anamika Publishers, 2003), p. 73.

63. D. Kaushik, "A Decade of Independent Tajikistan: Some Reflections," pp. 39–40; emphasis added.

64. K. Warikoo, "The Afghanistan Crisis and Tajikistan," in M. Singh (ed.), *India and Tajikistan: Revitalizing a Traditional Relationship* (New Delhi: Anamika Publishers, 2003), p. 151; emphasis added.

65. S. K. Asopa, "India and Post-Soviet Asia: An Appraisal of India's Central Asian Policy," p. 173.

66. R. Bedi, "India and Central Asia: India's Growing Presence and Role in the Resource-rich and Strategically Significant Central Asian Region Has Important Implications," *Frontline*, vol. 19, no. 19, September 14–27, 2002.

67. M. Singh, "India and Tajikistan: A Perspective for the 21st Century," in M. Singh (ed.), *India and Tajikistan: Revitalizing a Traditional Relationship* (New Delhi: Anamika Publishers, 2003), pp. 201–202.

68. R. Pandit, "Indian Forces Get Foothold in Central Asia," *The Times of India*, July 18, 2007.

69. C. R. Mohan, "Indian Foreign Policy," *World Focus*, vol. 28, no. 335/336 (2007), p. 395; emphasis added.

70. C. R. Mohan, *Crossing the Rubicon: The Shaping of India's New Foreign Policy*, p. 80.

71. S. Mansignh, "Why China Matters," in K. Bajpai, A. Mattoo (eds.), *The Peacock and the Dragon: India-China Relations in the Twenty-first Century* (New Delhi: Har-Anand Publications, 2000), p. 159.

72. B. M. Jain, *Global Power: India's Foreign Policy, 1947–2006*, p. 137.

73. M. S. Bhattacharjea, "Does China Have a Grand Strategy?" in P. Vohra, P. K. Ghosh (eds.), *China and the Indian Ocean Region* (New Delhi: National Maritime Foundation, 2008), p. 2.

74. U. Kachru, *Extreme Turbulence: India at the Crossroads*, p. 97.

75. S. Dutta, "China's Emerging Ties with Central Asia," in N. Joshi (ed.), *Central Asia—the Great Game Replayed: An Indian Perspective* (New Delhi: New Century Publications, 2003), p. 159.

76. R. G. Gidadhubli, "Politics of Oil and Natural Gas in Central Asia: Conflicts and Cooperation," p. 170.

77. M. S. Bhattacharjea, "Does China Have a Grand Strategy?" p. 10.

78. S. Varadarajan, "Central Asia: China and Russia up the Ante," *The Hindu*, July 8, 2005; emphasis added.

79. M. Joshi, "India and the Future of Asia: Arranging a Soft Landing for Pakistan," in R. R. Sharma (ed.), *India and Emerging Asia* (New Delhi: Sage, 2005), p. 103.

80. V. Sahni, "A Continent Becomes a Region: Future Asian Security Architectures," p. 84.

81. S. Singh, *China-South Asia: Issues, Equations, Policies* (New Delhi: Lancer's Books, 2003), p. 325.

82. M. S. Bhattacharjea, "Does China Have a Grand Strategy?" p. 16.

83. M. Joshi, "Issues in Central Asian Stability," p. 4.

84. A. Patnaik, "Central Asia's Security: The Asian Dimension," in R. R. Sharma (ed.), *India and Emerging Asia* (New Delhi: Sage, 2005), p. 216.

85. S. K. Asopa, "India and Post-Soviet Asia: An Appraisal of India's Central Asian Policy," p. 180.

86. J. Bakshi, *Russia and India: From Ideology to Geopolitics* (New Delhi: Dev Publications, 1999), p. 282.

87. See E. Kavalski, "Venus and the Porcupine: Assessing the EU-India Strategic Partnership," *South Asian Survey*, vol. 15, no. 1 (2008), pp. 63–81.

88. S. Singh, *China-South Asia: Issues, Equations, Policies,* pp. 161–162.

89. K. S. Rana, *Asian Diplomacy: The Foreign Ministries of China, India, Japan, Singapore, and Thailand,* p. 163.

90. S. Mansingh, C. V. Ranganathan, "Approaches to State Sovereignty," in G. P. Deshpande and A. Acharya (eds.), *Crossing a Bridge of Dreams: Fifty Years of India and China* (New Delhi: Tulika, 2001), pp. 465–466.

91. K. R. Reddy, "India and China Rapprochement: Cautious Moves," in Y. Y. Reddy (ed.), *Emerging India in Asia-Pacific* (New Delhi: New Century Publications, 2007), p. 212.

92. S. Dutta, "China's Emerging Ties with Central Asia," p. 155; emphasis added.

93. S. Singh, *China-South Asia: Issues, Equations, Policies,* p. 331.

94. S. Ganguly, "India and China: Border Issues, Domestic Integration, and International Security," in F. R. Frankel and H. Harding (eds.), *The India-China Relationship: What the United States Needs to Know* (New York: Columbia University Press, 2004), p. 124.

95. R. Budania, "The Emerging International Security System: Threats, Challenges, and Opportunities for India," in *Strategic Analysis*, vol. 27, no. 1 (2003), p. 89; emphasis added.

96. A. Mattoo, "USA and Central Asia: Beginning of the Great Game," in N. Joshi (ed.), *Central Asia—the Great Game Replayed: An Indian Perspective* (New Delhi: New Century Publications, 2003), p. 64.

97. V. C. Khanna, "Implications for India: Competition or Cooperation?" in P. Vohra, P. K. Ghosh (eds.), *China and the Indian Ocean Region* (New Delhi: National Maritime Foundation, 2008), p. 95.

98. R. G. Gidadhubli, "Politics of Oil and Natural Gas in Central Asia: Conflicts and Cooperation," p. 169.

99. S. Singh, *China-South Asia: Issues, Equations, Policies*, p. 323.

100. K. R. Reddy, "India and China Rapprochement: Cautious Moves," p. 201.

101. A. Patnaik, "Central Asia's Security: The Asian Dimension," p. 215.

102. S. Singh, *China-South Asia: Issues, Equations, Policies*, p. 107.

103. V. C. Khanna, "Implications for India: Competition or Cooperation?" p. 83.

104. A. Mattoo, "USA and Central Asia: Beginning of the Great Game," p. 20.

105. M. K. Bhadrakumar, "India Must Return to the Eurasian Energy Game," *The Hindu*, February 18, 2008.

106. K. S. Rana, *Asian Diplomacy: The Foreign Ministries of China, India, Japan, Singapore, and Thailand*, pp. 33–44.

107. M. S. Bhattacharjea, "Does China Have a Grand Strategy?" pp. 3–4.

108. S. Mansingh, C. V. Ranganathan, "Approaches to State Sovereignty," pp. 464–465.

109. A. Mattoo, "Imagining China," in K. Bajpai, A. Mattoo (eds.), *The Peacock and the Dragon: India-China Relations in the Twenty-first Century* (New Delhi: Har-Anand Publications, 2000), pp. 20–21.

110. E. Kavalski, "Partnership or Rivalry between the EU, China and India in Central Asia"; E. Kavalski, "The Fifth Debate and the Emergence of Complex International Relations Theory," *Cambridge Review of International Affairs*, vol. 20, no. 3 (2007), pp. 435–454.

111. S. Dutt, *India in a Globalized World* (Manchester: Manchester University Press, 2006), p. 205; emphasis added.

112. S. K. Mitra, "The Reluctant Hegemon: India's Self-perception and the South Asian Strategic Environment," p. 400.

113. S. Singh, *China-South Asia: Issues, Equations, Policies*, p. 323.

114. J. E. C. Hymans, "India's Soft Power and Vulnerability," *India Review*, vol. 8, no. 3 (2009), p. 259; S. K. Mitra, "The Reluctant Hegemon: India's Self-perception and the South Asian Strategic Environment," p. 404.

115. B. M. Jain, *Global Power: India's Foreign Policy, 1947–2006*, p. 228.

116. P. C. Alexander, *Through the Corridors of Power: An Insider's Story* (New Delhi: HarperCollins, 2004), p. 357.

117. M. K. Bhadrakumar, "India Must Return to the Eurasian Energy Game."

118. See P. B. Mehta, "Still under Nehru's Shadow? The Absence of Foreign Policy Frameworks in India."

119. D. Lak, *The Future of a New Superpower* (New York: Viking, 2008), p. 260; emphasis added.

120. V. Rai, W. Simon, *Think India* (New York: Dutton, 2007), p. ix.

121. N. Menon, A. Nigam, *Power and Contestation: India since 1989*, p. 85.

122. G. Das, *India Unbound* (New Delhi: Viking, 2000).

123. P. B. Mehta, "Still under Nehru's Shadow? The Absence of Foreign Policy Frameworks in India."

124. R. Kumar, "India as a Foreign Policy Actor: Normative Redux," p. 255.

125. R. W. Jones, *India's Strategic Culture*, p. 21; S. K. Mitra, "The Reluctant Hegemon: India's Self-perception and the South Asian Strategic Environment," p. 402.

126. M. S. Bhattacharjea, "Does China Have a Grand Strategy?" p. 10.

127. U. Kachru, *Extreme Turbulence: India at the Crossroads*, p. 14.

CHAPTER 5

Afghanistan and Regional Strategy: The India Factor

MEENA SINGH ROY[1]

Afghanistan—a plural, multiethnic, and multilingual country—has been an area of grand ambitions and competition for imperial powers both in medieval and modern history. Despite foreign interventions and repeated violence between various power structures, Afghanistan has been able to continue as a geopolitical unit. Its geographical location has always attracted the attention of regional and extra-regional powers. Afghanistan has been the playground for these powers either to retain their influence or to contain their adversary.[2] These competitions and rivalry among various actors have negatively effected the political and economic development of Afghanistan. However, this is not to say that other factors have not contributed toward the instability and tensions within the country. The domestic conflict and competition between the forces of modernization and orthodox Islam and the ethnic divisions within the Afghan population have also afflicted Afghanistan.[3]

The only period that could perhaps be considered comparatively stable was from 1933 to 1973 when King Mohammed Zahir Shah ruled the country. The political process and structure of the government during this period remained decentralized and feudal. King Zahir Shah's developmental activities focused mainly on the capital (Kabul) and other important cities. Rural Afghanistan was left to its tribal leaders to manage their affairs according to the local traditional practices and customs. In 1973, Afghanistan became a battleground for Daud's bloodless coup against King Zahir Shah followed by the Soviet-backed violent revolution resulting in Soviet military intervention in 1979 and the ensuing cold war politics between the United States and the Soviet Union.[4] The situation deteriorated in Afghanistan after the Soviet withdrawal in 1989, leaving the country into the hands of the Taliban. During the years of Taliban control, Afghanistan became a hotbed of extremism and a breeding ground for jihadi forces involved in exporting extremism and terrorism in the region and beyond. In the post-9/11 era, the U.S. intervention in Afghanistan was seen by many as a stabilizing force in war-torn Afghanistan. Today, the security situation in Afghanistan is worrisome. Increasing violence, a resurgent Taliban, rampant drug trafficking, the Karzai government's inability to deliver, and increasing local discontent have all become a serious cause of concern for the international community, which has been struggling to bring some stability in Afghanistan since 2001.[5] Moreover, the deteriorating situation in Pakistan has further complicated the situation in the region.

President Obama's new Af-Pak Strategy and his subsequent commitment of additional 30,000 troops and setting of a timetable for a "draw down" of forces have brought to the forefront the critical situation that is still unfolding in this region. In the current context, various options to stabilize Afghanistan are under consideration in a number of regional, subregional, and international mechanisms. Some of the suggested options are (1) use of diplomatic tools to engage Taliban; (2) putting greater emphasis on multilateral initiatives; (3) strengthening security; (4) promoting economic growth; (5) strengthening institutions; (6) and, more importantly, enhancing regional cooperation. As the NATO and U.S. forces have been unable to put down the insurgency in Afghanistan after the U.S.-led coalition ousted the Taliban from power, there seems to be a general consensus that regional countries need to play a greater role in Afghanistan. From Obama's Af-Pak strategy to the London Conference on Afghanistan, importance of regional cooperation in addressing the Afghan issue has been highlighted. Speaking on the same issue recently during a major security conference in Munich, the head of NATO, Anders Fogh Rasmussen, pointed out that it is vital to boost ties with countries like India, China, and Pakistan. He said that "we need an entirely new compact between all the actors on the security stage" to meet current security requirements in Afghanistan. Although there is a consensus that regional cooperation is important to rebuild Afghanistan, it is not very clear what kind of mechanism will work as every country has its own vested interest in Afghanistan. It is equally important to examine how effective the regional mechanisms have been in addressing the Afghan quagmire? It is in this context that the present study examines the Afghan situation with a special focus on the role of regional actors, particularly India, and the challenges and options in Afghanistan. The present study seeks answers to two important questions:

i. Is there a regional strategy capable of solving the Afghan quagmire? How are regional countries responding to the situation in Afghanistan?

ii. Can India play any significant role in Afghanistan under the current situation? If yes, what are its scope and limitations?

Afghanistan: Strategic Quagmire

Regional and Subregional Dynamics

The regional dynamics in Afghanistan is much more complex and nuanced than what it appears to be. Today, Afghanistan is exposed to a completely new set of challenges. Situation within Afghanistan is mired in the geopolitics of regional and extra-regional players with global implications. In March 2009, the Afghan foreign minister pointed out that terrorism and narcotic drugs[6] followed by the fragility of state institutions and socioeconomic challenges such as unemployment and poverty were the main threats faced by his country.[7] Afghanistan continues to be closely entangled with the overall security of the region. Developments in Afghanistan have serious security implications for its neighborhood—Central Asia, West Asia, and South Asia. The Afghan quagmire needs to be viewed in the context of regional and subregional dynamics that is still unfolding. Instability in Afghanistan has spillover effect on Central Asian Republics, India, Iran, Russia, and China.

Afghanistan is an issue of common concern for all the Central Asian leaders. Tajikistan, Uzbekistan, and Turkmenistan share borders with Afghanistan. Tajikistan is particularly vulnerable as it shares 1,340 kilometers of borders with Afghanistan. The Central Asian countries have developed close bilateral relations with Afghanistan. Uzbekistan, Tajikistan, and Kyrgyzstan, though worst affected by instability in Afghanistan, have been able to do very little to address the problems due to their limited military and economic resources and have been dependent on powers such as Russia, China, and the United States to find solutions to the problems emanating from Afghanistan and Pakistan. They have been very vocal in the Shanghai Cooperation Organisation (SCO) to address the Afghan issue. Uzbekistan, as a part of its effort to address the Afghan challenge, has proposed a reorganizing of the UN-sponsored 6 + 2, which now includes the NATO but does not include Afghanistan, into a 6 + 3. Uzbek president Islam Karimov has expressed concern over the militarization of Afghanistan; he has urged the international community to focus more on the resolution of his country's social and economic issues. Expressing his concern about the problems in Afghanistan he said, "We are alarmed by the situation in Afghanistan and growth in drug production and trafficking. Stability in Afghanistan has to be found in the resolution of internal social issues rather than further militarization of the country."[8]

Tajikistan's security is closely linked to the security of Afghanistan. As an immediate neighbor, Tajikistan, during the civil war that broke out in the initial years of its independence, has experienced the spillover effect of instability in Afghanistan. Since the beginning of 2010, Kunduz province bordering Tajikistan has become the hotbed of conflict between the

International Security Assistance Force (ISAF) and Taliban insurgents. The increase in Taliban operations in this province is attributed to the newly activated Northern Distribution Network, an overland supply route for the ISAF that connects Western Europe to Afghanistan via Central Asian States.[9] Tajikistan has been seeking more economic and military assistance to deal with the narcotic problem. The Tajik foreign minister Hamrokhon Zarifi has articulated the view that the problem within Afghanistan cannot be settled by military means or by force. According to him, "socio-economic revival was and remains the most efficient factor" to bring about changes in Afghanistan. Tajikistan has proposed stringent measures to ensure strict control of borders with neighboring countries to control the circulation of illegal drugs, a major threat confronting every country in the region and beyond. It is believed that the implementation of the program to sow alternative agricultural crops may become an important factor in eliminating cultivation of opium. The Tajik president has proposed to create a third center for fighting circulation of illegal drugs in Dushanbe.[10] Tajikistan has also expressed its willingness to provide Afghan transport operators with transit to China. In the long term, Tajikistan is also ready to connect Afghanistan's railroads to those in Tajikistan, thus helping economic integration in the region. It has also offered to work jointly with Afghanistan on constructing hydroelectric power stations, which will help irrigate millions of hectares of land in northern Afghanistan.[11]

Kazakhstan does not share land borders with Afghanistan and, therefore, is not directly affected by developments in Afghanistan. But because of its borders with Uzbekistan and Kyrgyzstan it wants to control the inflow of terrorists and narcotics. It has been providing Afghanistan aid and help. In 2007, the Kazakh president announced that his country would be involved in humanitarian projects in Afghanistan. Kazakhstan has been running special programs supporting Afghanistan in training graduates in various areas, in the construction of infrastructural facilities, and in rendering humanitarian aid. Kazakh businessmen have shown serious interest in making investment in mutually beneficial projects in Afghanistan.[12] Kazakhstan will allocate US$50 million for Afghan–Kazakh cooperation in the education sector. Starting from 2010, Kazakhstan will train 1,000 Afghan specialists for the next five years in the higher educational centers and vocational schools in Kazakhstan.[13] On the issue of addressing the current challenges in war-torn Afghanistan, Kazakhstan has emphasized the importance of both multilateral and bilateral cooperation. Kazakh first deputy foreign minister Nurtay Abykayev has suggested that in the UN, NATO and such structures as the Organisation of the Islamic Conference (OIC) and the Economic Cooperation Organisation (ECO) should be fully involved in the resolution of the Afghan issue and has highlighted the need to create reliable "zones of anti-terrorist, anti-drug and financial security" around Afghanistan.[14]

Pakistan's desire to expand its influence in Afghanistan has been very central to Islamabad's strategic thinking. In the view of Gen. Mirza Aslam Beg, "Afghanistan will be a great source of strength to Pakistan to face any crisis and danger that it may encounter. In fact, the security of Pakistan, Iran and Afghanistan is interlinked and is indivisible."[15] While examining Pakistan's Afghan policy, it has been argued that Pakistan's defense planning has been handicapped by the lack of territorial depth to absorb an attack by India and then to retaliate. This elusive quest for strategic depth has guided Pakistan in its ambitious involvement in Afghanistan in spite of multidimensional implications for its social fabric and political culture. Thus, Pakistan's Afghan policy was constructed with an objective to create a subservient government in Afghanistan that would be friendly to Pakistan, militarily too weak to question the Durand line, and politically too unstable to raise the Pashtunistan issue. In addition to this, the military strategists argued that a friendly Afghanistan would give Kashmiri militants a base from where they could be trained, funded, and armed.[16] Pakistan views Pashtun nationalism as an existential threat, and its 2,500-kilometer-long disputed border with Afghanistan has been at the core of Islamabad's policy of interference in Afghanistan's internal affairs. Despite Islamabad's repeated claims of noninterference, its military support continues for Afghan Taliban and Islamist terrorist groups operating against India.[17] The crisis in Afghanistan needs to be viewed in the context of Islamabad's reliance on Islamist non-state actors as an instrument of its state policy. In fact, Pakistan began its adventurism with the Islamist forces in Afghanistan way back in 1973. The then prime minister, Zulfikar Ali Bhutto, provided sanctuary to Islamist leader Gulbuddin Hekmatyar with a view to undermine the established government in Kabul. This was six years before Soviet intervention in Afghanistan.[18] The end of the Taliban regime in Afghanistan has not brought any fundamental change in Pakistan's Afghan policy. Afghanistan is still very central to Pakistan's strategic thinking and is akin to being Pakistan's backwaters. Wary of how the current situation would unfold in Afghanistan, Islamabad is trying to cement its ties with the current regime in Kabul while also supporting Taliban elements. It has not given up its grand strategy of establishing a friendly government in Kabul and denying any role to India in Afghanistan.

For Russia, Afghanistan is crucial for peace and security in its backyard in Central Asia. Moscow is concerned about the increasing drug trafficking in the region and insurgency in Chechnya. The linkage between the Chechen rebels and extremist forces in Afghanistan has serious security implications for Russia. During the March 2009 Moscow conference on Afghanistan, the Russian foreign minister, Sergey Lavrov, said that drug trafficking has become the most serious threat for Russia and the countries of Central Asia. According to the Federal Drug Control Agency, Russia is the world's biggest heroin consumer and has been alarmed by a surge in trafficking of the opiate through Central Asia and across its territory. Twelve tons of pure heroin, enough for about 3 billion single doses, arrives in Russia each year from Afghanistan. It was argued that to "stabilize the situation a comprehensive approach is needed which combines the military suppression of terrorists, extremists and drug dealers with wide-scale programme of economic and social rehabilitation. It is important to win the trust and support of the entire Afghan people."[19] Russia has offered to cooperate with the regional powers and U.S.-led forces fighting in Afghanistan and to take active joint steps aimed at normalizing the situation in Afghanistan.[20] Russia has allowed shipment of military cargoes through its territory to support ISAF activities in Afghanistan. It has also promised to resume defense supplies to Afghanistan at a meeting of Russia-U.S. Working Group on Counterterrorism in Moscow in June 2008. Russia supplied US$220 million worth of military equipment to the Afghan army during 2002–2005 but had halted the supply later.[21] At the same time, Moscow has expressed its opposition to the expansion of U.S. military bases in Central Asia and to NATO's eastward expansion at the same time. Despite Russia's friendly gestures, the aforesaid issues of NATO's expansion and the U.S. military expansion in CARs (Central Asian Republics) are likely to remain an irritant between the United States and Russia. It is important to note that despite its support to stabilize Afghanistan, Moscow has made it very clear that it has no intention of sending a military contingent. Russian foreign minister Lavrov during his visit in March 2009 to Kabul stated that "the international community is not asking the Russian Federation to send its military contingent to Afghanistan, nor is any such action being planned."[22] Russia's role remains crucial in Afghanistan.

China has been concerned about the unstable political situation within Pakistan and Afghanistan because of Taliban's ties with Islamic extremist groups in the Xinjiang region in Western China. Moreover, in post-Taliban period China has become a major economic stakeholder in Afghanistan. In recent years, Beijing has adopted a policy of bilateral engagement with Kabul to pursue economic diplomacy. It has relied more on Pakistan to deal with the Taliban problem because of its close strategic ties with Islamabad. Since the fall of the Taliban regime in Afghanistan, China has been involved in the reconstruction of that country by providing aid and assistance, which has been in limited supply. In 2006, China promised a total of US$10 million as assistance and also agreed to abolish tariffs on Afghanistan's exports to China.[23] Some Chinese companies have been involved in Afghanistan. The ZTE and Huawei are partners with the Afghan Ministry of Communications to install digital telephone switches and are providing some 200,000 subscriber lines.[24] In restoring water supply in Parwar Province, China participated in the Parwan irrigation project as well. In addition, it has helped Afghanistan in the reconstruction of a public hospital in Kabul.[25] In May 2008, China acquired major economic stakes in Afghanistan winning a US$3.5 billion contract to develop Afghanistan's huge Aynak copper field. This contract also involves construction of a power plant and a railroad connecting the mines to China through Pakistan. China is also using the regional mechanism of SCO but in a much selected manner. It is also interested in Afghanistan's unexplored reserves of oil and natural gas in its northern area. China is unlikely to follow any activist approach in Afghanistan as long as NATO and American forces are present there.

For Iran, Afghanistan is of great strategic significance. Since 2001, the Iranian government has engaged in cordial relations with Afghanistan. Despite its strained relations with the United States, it has enhanced its trade and economic relations with Afghanistan. It has built roads, power transmission lines, and border stations in Afghanistan. Iran has influence in the northwest part of Afghanistan. More importantly, Iran has been host to more than 2 million Afghan refugees during the last two decades. According to UNHCR's estimates there are still some 950,000 Afghan refugees in Iran. Resurgent Taliban is a serious security concern for Iran. Instability in Afghanistan will have an adverse effect on Iran's security in areas bordering Afghanistan. Soon after suicide bombers carried out a horrific attack in October 2009 on Iran's Revolutionary Guards in southeast of Iran, the Pakistan-based Jundallah, a Sunni extremist outfit, claimed responsibility for it. This was the first time terror outfits based in Pakistan targeted Iranian territory.[26] Tehran is opposed to the Taliban and had backed the Northern Alliance and supported the United States in the ouster of the Taliban regime in 2001. Despite its support to the United States, it is wary of U.S. military presence in Afghanistan. The increasing violence and instability in Pakistan and constant attacks on NATO's military supply routes in Pakistan enhance the role of Iran in providing safe routes for the ISAF supplies for Afghanistan. Iran's ability to play a significant role in dealing with the Afghan quagmire, particularly in northwestern Afghanistan, cannot be ignored. Given Iran's interests in Afghanistan, it will continue to cultivate its links with Hazaras, Tajiks, and Uzbeks, who they have supported in the past. Given Iran's influence in northeastern Afghanistan, it is important for the United States and Europe to engage Tehran in dialogue on the Afghan issue.

There are some subregional mechanisms at work to address the Afghan issue. In the past, India, Russia, and Iran together with Tajikistan have worked to support the Northern Alliance against Taliban. In the current context, Russia, Pakistan, and Tajikistan together with Afghanistan have met to coordinate activities of regional states in the battle against terrorism and extremism and in the promotion of intra-regional cooperation in trade and development. On the sidelines of the SCO Summit in Yekaterinburg on June 15, 2009, the presidents of Pakistan, Afghanistan, and Russia held, for the first time, a trilateral meeting.[27] In January 2010, Afghanistan, Iran, and Pakistan signed an agreement to prepare a joint framework to address the problem of terrorism and extremism. In a trilateral meeting, the foreign ministers of the three countries acknowledged that "terrorism poses a common challenge that can only be addressed through concerted efforts."[28] At the same time, Turkey, Pakistan, and Afghanistan are also working toward creating a regional mechanism to address the Afghan quagmire and rebuild war-torn Afghanistan.

There have been three summits between Turkey, Pakistan, and Afghanistan. The Istanbul conference was one such subregional initiative of Turkey to focus on the issue of security in Afghanistan and to facilitate closer cooperation between Pakistan and Afghanistan. Although India is a major contributor to Afghanistan's nation-building efforts, it was kept out of this meeting. President Karzai has requested repeatedly that Saudi Arabia play the role of a mediator between the Taliban and his government. Saudi officials have accepted the request but only if the Taliban breaks all ties with Al Qaeda. However, the Taliban is unlikely to break its links with Al Qaeda. This leaves little room for Saudi mediation.[29] In addition, there are several regional groupings—such as the South Asian Association for Regional Cooperation (SAARC), the Regional Economic Cooperation Conference on Afghanistan (RECCA), the SCO, the Economic Cooperation Organisation (ECO), and the Organisation of the Islamic Conference (OIC)—that are deeply concerned about developments in Afghanistan and have been trying to address the Afghan issue in these regional organizations. However, the political rivalries and trust deficit among the regional countries, their economic limitations, and their broader strategic agreement on the future of Afghanistan undermine efforts to promote cooperation among the regional countries. The political differences between the regional and extra-regional actors present in Afghanistan—the U.S.-Iran stand off, lack of trust between the United States and Russia, and between the United States and China—further complicate the situation. Despite their support to the U.S. war in Afghanistan after 9/11, both China and Russia are wary of American long-term military presence in the region. The United States is focused more on dealing with Iran's nuclear issue than on engaging Tehran to play a positive role in Afghanistan.

Afghanistan: The India Factor

Afghanistan in India's Strategic Calculus

The analysis of regional dynamics in Afghanistan would remain incomplete without examining India's role in Afghanistan. Afghanistan is at the center of India's strategic calculus. As the regional power on the Indian subcontinent, India has legitimate strategic interest in the stability and security of Afghanistan. A stable Afghanistan is in India's interest. Instability in Afghanistan is detrimental to India's security. In the past, Pakistan's intelligence agency used Afghanistan to train terrorists and export jihadis to the Jammu and Kashmir region of India. In the current context, Pakistan-backed Taliban militants continue to hamper infrastructure development work being carried out by India in Afghanistan. From July 2008 to February 2010, there have been three direct attacks on Indians in Afghanistan. The Pakistan-backed Taliban were blamed for the attacks on the Indian embassy in Kabul in July 2008. In October 2009, another attack claimed the lives of Indian diplomats and officials based in Kabul and many Afghan people. The latest being the February 26, 2010 attack on a hotel and guesthouse full of Indian doctors, engineers, and security personnel; it claimed at least nine Indian lives. The Taliban has claimed responsibility for the attack. This is an attempt to undermine increasing Indo-Afghan ties. India is the biggest partner in Afghan reconstruction and other developmental projects being carried out by hundreds of Indians working there.[30]

India's growing ties with the Karzai government and its increasing role in rebuilding Afghanistan have been viewed by Islamabad with suspicion. It has accused India of backing the Baluchistan insurgency in Pakistan. In July 2009, Pakistan's interior minister, Rehman Malik, stated that Islamabad has enough proof that India and Afghanistan are involved in the ongoing unrest in Baluchistan. These allegations have been refuted by both India and Afghanistan. The Afghan foreign minister during his visit to India in July 2009 stated that Afghanistan has never allowed any country to interfere in the domestic issues of Pakistan and that India has never used Afghan territory against Pakistan.[31]

Afghanistan's importance for India lies in its strategic location. It is a gateway for India to resource the rich Central Asian region. A stable Afghanistan can provide access to oil and gas in Iran, Central Asia, and the Caspian region. India is also seeking to have friendly ties with countries in its neighborhood to ensure a stable and secure regional environment to enhance economic and trade relations with them. It does not want the region to become a hotbed of extremist forces or a springboard for insurgencies against India.[32]

India's Engagement: Soft Power Strategy

The Indo-Afghan relationship goes back centuries. In the past, much before British imperial power came to the region, India had extensive cultural and trade links with Afghanistan. India maintained close ties with King Zahir Shah of Afghanistan; even after his ouster in 1973, New Delhi managed to keep close ties with subsequent governments. India's ties with Kabul ended after Taliban seized power in 1996 only to be reestablished in 2001 after American-led forces dismantled the Taliban regime. However, during the Taliban control of Afghanistan India supported the Northern Alliance and provided them military and logistic backing.[33]

Since 2001, India has adopted a soft power strategy in Afghanistan. Its policy has been that of providing humanitarian assistance, capacity building, augmenting economic growth, developing infrastructure, and working toward integrating Afghanistan into the South Asian cooperative framework with an aim of reviving the traditional role of Afghanistan as a land bridge connecting South Asia with Central Asia and West Asia. Today Afghanistan is a SAARC member country.

India has always seen Afghanistan as a land bridge of trade. This view was expressed in the Moscow Conference by the prime minister's special envoy, S. K. Lambah. He said that "historically, Afghanistan has prospered when it has served as the trade and transportation hub between Central Asia and South Asia. If we were to implement the projects and activities on the anvil, which allow greater commercial and economic exchanges by removing barriers to investment, trade and transit, this would transform not just Afghanistan but other regional countries as well."[34]

India has been trying to work toward realizing this goal. Its strong economy has contributed in a major way to the reconstruction of Afghanistan and has the potential to continue doing so. However, in the past few years its activities have been hampered due to the deteriorating security situation in Afghanistan and the constant effort of extremist groups backed by Pakistan to sabotage India's reconstruction work. Following the Kabul Declaration of December 2005, the Second Regional Economic Cooperation Conference was hosted jointly by India and Afghanistan in New Delhi in November 2006. Many of the decisions taken during that conference have yet to be implemented. New Delhi has emphasized the importance of regional cooperation because this can help in addressing transborder issues, developing commercial and economic opportunities, and ending cross-border infiltration and terrorism. India has also built the 218-kilometer-long Zaranj-Delaram Road at a cost of US$150 million (Rs. 750 crore), which was inaugurated by External Affairs Minister Pranab Mukherjee in January 2009. This will help link Afghanistan with Chahbahar Port in Iran. The Pul-e-Khumri transmission line to Kabul will soon be completed by the Power Grid Corporation of India. This transmission line brings in power from Uzbekistan to Kabul and should be seen as India's attempt to connect Afghanistan with its neighbors and make them stakeholders in Afghanistan's development. The government-run Water and Power Consultancy Services Ltd. (or WAPCOS) is in charge of the largest project that India has undertaken—the 42-MW Salma Dam Power Project in the western Afghanistan province of Herat.[35]

India's pledged bilateral commitment to the rebuilding and reconstruction of Afghanistan is now worth US$1.2 billion. In January 2009, India also announced an aid of 250,000 metric tons of wheat to Afghanistan to help tide over its food crisis.[36] There are over 4,000 Indians working in Afghanistan on various projects in institution-building, infrastructure, education, power, telecommunications, agriculture, and food assistance. There are 43 registered Indian and Indian joint-venture companies in Afghanistan.[37] India has also provided assistance to Afghanistan in education, health care, and training of Afghan diplomats and police. India's trade with Afghanistan has gone up over the last five years moving from US$212.44 million in 2004–2005 to US$520.47 million in 2008–2009 (see table 5.1).

Indo-Afghan ties got new inputs during Afghan external affairs minister Rangin Dadfar Spanta's two-day visit to Delhi in July 2009, when both the countries decided to set up an India-Afghanistan Partnership Council composed of separate groups on political consultation, capacity development and education, power and water, culture, trade and industry, health, and agriculture. This is an attempt to enlarge developmental cooperation and harness greater institutional support for the implementation of ongoing developmental programs.[38]

Table 5.1 India-Afghanistan Trade (values in US$ million)

Per Year	2004–2005	2005–2006	2006–2007	2007–2008	2008–2009
EXPORT	165.44	142.67	182.11	249.21	394.23
Growth percentage		−13.76	27.64	36.85	58.20
India's total export	83,535.94	103,090.53	126,414.05	163,132.18	185,295.36
Growth percentage		23.41	22.62	29.05	13.59
Share percentage	0.20	0.14	0.14	0.15	0.21
IMPORT	47.01	58.42	34.37	109.97	126.24
Growth percentage		24.29	−41.16	219.92	14.80
India's total import	111,517.43	149,165.73	185,735.24	251,654.01	303,696.31
Growth percentage		33.76	24.52	35.49	20.68
Share percentage	0.04	0.04	0.02	0.04	0.04
TOTAL TRADE	212.44	201.09	216.48	359.18	520.47
Growth percentage		−5.34	7.65	65.92	44.91
India's total trade	195,053.37	252,256.26	312,149.29	414,786.19	488,991.67
Growth percentage		29.33	23.74	32.88	17.89
Share percentage	0.11	0.08	0.07	0.09	0.11
TRADE BALANCE	118.43	84.24	147.73	139.24	268.00
India's trade balance	−27,981.49	−46,075.20	−59,321.19	−88,521.83	−118,400.95

Note: The country's total imports (S.No.6) since 2000–2001 does not include import of Petroleum Products (27100093) and Crude Oil (27090000)

Source: Department of Commerce, Government of India at http://commerce.nic.in/eidb/iecnt.asp (accessed on Feburary16, 2010)

India has expressed its support for the continued international attention on Afghanistan's rehabilitation and security. In July 2009, India's external affairs Minister, S. M. Krishna, said that India has a direct interest in the success of the international efforts in stabilizing Asia and added that India was playing a substantial role in the reconstruction and assistance of Afghanistan.[39] The current status of relationship between the two countries is a clear indication of India's sincere interest in rebuilding Afghanistan through its soft power strategy. Indian leadership has repeatedly expressed its support for the aspirations of Afghan people to build a peaceful, prosperous, democratic, and pluralistic nation.[40] India's Afghan policy has been that of keeping out of the great power politics and actively supporting the regime in Kabul through its soft power strategy. India cannot remain indifferent to the developments in Afghanistan because it has serious security implications for India.

Challenges for India

India's extensive involvement in Afghanistan to stabilize that country has been appreciated by the European Union, Russia, Iran, and Afghanistan but its greatest challenge comes from Pakistan and its support to Islamist groups in Afghanistan. As mentioned earlier, Pakistan's Afghan policy is aimed at containing India's influence in Afghanistan. Unfortunately, Pakistan does not perceive any role for India in Afghanistan. Pakistani officials still regard India as their strategic priority. In a July 2009 briefing, a Pakistani official clearly pointed out that, however concerned the United States may be about the Taliban, Pakistan still regarded India as their top strategic priority and the Taliban militants as little more than a containable nuisance and as potential long-term allies.[41] In fact, Pakistan's foreign policy, since its very inception, has been conditioned by two interrelated factors: the fear of India and an urge to seek a strategic balance with India. These strands determine Islamabad's Afghan policy as well.[42] General Ashfaq Parvez Kayani, Pakistan's army chief, has recently said that his army was India–centric and, therefore, entitled to "Strategic depth" in Afghanistan.[43] Pakistan's desire to have strategic depth in Afghanistan against India and its continuous support to the Islamist forces against India to keep India out of Afghanistan not only complicate the situation in the region but also hamper any effort of developing meaningful regional cooperation to rebuild and stabilize Afghanistan.

In addition to Pakistan's suspicions about India's role in Afghanistan and Pakistani strategy to contain Indian influence in the region, President Obama's new Af-Pak strategy puts pressure on India's role in Afghanistan. The new Af-Pak strategy does not acknowledge India's role and is focused on Pakistan and its role in resolving the Taliban issue in Afghanistan. Pakistan has once again become the fulcrum for U.S strategy in Afghanistan. The White House, State Department, and Pentagon are seeking greater cooperation from Pakistan in addressing the Afghan crisis. The U.S strategy of greater reliance on Pakistan and its Intelligence agency ISI to strike a deal with the Taliban is further complicating the situation for India. Islamabad can leverage its newly acquired importance to secure its own interests vis-à-vis India in Afghanistan. The October 2009 confidential report by General Stanley McChrystal, U.S. top commander in Afghanistan, acknowledged that "Indian activities largely benefit the Afghan people" but at the same time pointed out "increasing Indian influence in Afghanistan is likely to exacerbate regional tensions and encourage Pakistani counter measure in Afghanistan or India." Such thinking in turn works well with Pakistan's fear that India's presence in Afghanistan is intended to encircle Pakistan and thus serves as justification for Islamabad's continued support to the Taliban.[44]

From India's perspective, cutting any deal with the Afghan Taliban will be inimical to its interests. Indian analysts are of the view that "a stable Afghanistan cannot emerge without dismantling the Pakistani military's sanctuaries and sustenance infrastructure for the Afghan Taliban."[45] It is argued that even if the Obama administration is able to bring down the violence by striking a deal with the Taliban, the Taliban would still remain intact as a militant force with strong ties with the Pakistani army. This will have serious implication for regions inflicted with extremism and jihadi forces in South Asia and beyond. A realistic approach for the international community is to ensure that extremist forces and Al Qaeda are denied operating bases in Afghanistan and Pakistan.

The U.S. strategy in Afghanistan has not been able to bring about the desired stability. The violence is at its highest and the Taliban is much more active than before even in areas where it had no control earlier, thus increasing discontent at local level, and the government in Kabul is corrupt and inefficient; all these are indicating that the situation is grim. The president of the Council on Foreign Relations (CFR), Richard N. Haass, in his recent writings has stated that "the U.S. is now too stretched economically and militarily to succeed by relying solely on its own resources."[46] In the current context the United States is looking for quick-fix solutions to the Afghan quagmire by finding ways of accommodating the Taliban. President Obama's "renewed" commitment to resource the "just war" by agreeing to send 30,000 U.S. troops is unlikely to bring the desired result of reversing the tide of the Taliban momentum at least in a year's time. General McChrystal's counterinsurgency strategy of "clear, build, and transfer" is unlikely to deliver quick results. Under the current Afghan situation, which is still evolving, the American policymakers are confronted with tough policy choices. A premature exit of the U.S.-NATO forces from Afghanistan will not be in India's interest. India's developmental projects continue to be targeted by the Taliban-led insurgency. In the event of U.S. withdrawal or drawdown of forces, a return of the Taliban regime is not a farfetched possibility. The London conference on Afghanistan has evoked differing views in the Indian media. This conference is seen as a major setback for India as it calls for engaging the "good" Taliban and opens up the possibility of a Taliban regime in Kabul. India has never supported the idea of distinguishing between the "good" Taliban and the "bad" Taliban. Some Indian analysts have summed up the London conference in three words: "surge, bribe and run." It is argued that the intention of President Obama's troop surge is not to militarily rout the Afghan Taliban but to strike a political deal and buy them off.[47] Others are of the view that the London conference on the Afghan problem certainly gives grounds for optimism.[48] Current U.S Strategy to solve the Afghan quagmire has serious security implication for India.

Options for India: Tough Road Ahead

Given the current challenges that India faces in Afghanistan, its options are limited and the road ahead tough. After the United States' new Af-Pak strategy was declared, there is a line of thinking—at least in some sections of policymaking community and analysts—that India should downscale its presence in Afghanistan. At the same time there are others who are of the view that India's interests demand exploring the possibility of putting "boots on ground." However, this view of India's military involvement in Afghanistan does not find support in many quarters of either policymakers or analysts. It is argued that military involvement will not serve India's long-term strategic interests. Despite its limitations and problems in Afghanistan, India's possible options are the following:

- India should continue to support and strengthen the Karzai government with every possible financial and technical support. However, it should build its own constituency by engaging all sections in Afghanistan. It should develop contacts with Pashtun tribes particularly in southern and eastern Afghanistan.

- India's economic program and developmental projects have generated a lot of goodwill for India among the Afghan people. In the recently conducted Afghanistan opinion poll commissioned by the British Broadcasting Corporation, the ABC News, and Germany's ARD, India was rated highest: 29 percent of Afghans as "very favorable" with India, compared to other countries included in the survey: Iran—18 percent, Germany—17 percent, the United States—8 percent, Pakistan—2 percent. India was also rated highest (44 percent) as a country playing a "neutral" role.[49] Therefore, New Delhi should continue active involvement in capacity building and training Afghan nationals. It is argued that "India's priority should be to insulate national security from the negative fallout of the US-led war, while remaining focused on Afghan reconstruction."[50]

• India can help Afghan people in establishing small-scale industries at local level in carpet making, handicraft, ornament, and packaging of fruits and nuts. India can work under the SAARC mechanism for the economic integration of Afghanistan in the region.

• India can play an important role in augmenting and training the Afghan security and police forces. The Indian army has a vast experience in counter-insurgency operations in different terrains: the northeast, the high altitudes of Jammu and Kashmir, and the plains of Punjab. It can train the Afghan forces in Indian military institutions. India can share its experience with the Afghan government in building local institutions of governance.

• At a regional level Indian policymakers need to focus more on Central Asian countries, particularly Tajikistan, Turkmenistan, and Uzbekistan. Indian involvement in the various economic sectors of these countries should increase. It needs to invest in building roads and bridges to connect Central Asia with Afghanistan. India will have to pursue active economic diplomacy with these countries particularly Tajikistan.

• India will have to develop a new mechanism of cooperation with regional actors such as Russia and Iran on Afghanistan. There is ongoing cooperation with these countries but it needs to be intensified in the light of current developments in the region.

• The trilateral mechanism of India-Russia-China cooperation provides opportunity for stabilizing Afghanistan. In their recent meeting in October 2009 in India, they have agreed to jointly expand their cooperation in combating international terrorism. This mechanism needs to be strengthened. All the three countries have major stakes in a stable Afghanistan. In the current context, with increasing Chinese economic engagement China would not like to see Afghanistan return to pre-9/11 situation. Some Indian analysts have proposed the creation of a "Concert of Powers," a regional grouping including the United States, India, Iran, Central Asian states, and China, while others have argued for organizing a conference on "Afghanistan's Independence and Neutrality" at an international level.[51]

• Most importantly, India will have to find ways to engage Pakistan in dialogue despite Islamabad's fear about India's increasing influence in Afghanistan. India has always sought a policy of peaceful and normal relationship with Pakistan. Indian prime minister Manmohan Singh has said that there was no alternative to dialogue to resolve the Indo-Pak issue. He emphasized that "today the prime issue is terrorism. We are ready to discuss all issues with them in an atmosphere free from terrorism."[52]

• In the regional context, the SCO has been active in addressing the issue of Afghanistan. India remains an observer state and thus doesn't wield the same influence as Russia or China. However, the Afghan issue remains a common concern for all SCO member and observer states. Some Indian experts are of the view that "the SCO processes on the stabilization of Afghanistan serve India's interests."[53] It is argued that India needs to energize its moribund regional diplomacy by boosting its relationship with Iran and Russia as both these countries are major factors of regional stability. It is pointed out that India's outlook on the SCO must radically change. This is the organization that provides a useful forum to engage China and Pakistan in the issues of regional security.[54] At the same time other experts on the subject have argued that, "in absence of a direct land border with Central Asia, India's ability to assert itself in the SCO will be meager."[55] Given the current nature and functioning of the SCO it is in India's interest to continue with its observer status and boost its cooperation at bilateral level with the member states, particularly with Central Asian countries, Russia, and Iran. All these countries have commonality of interests in Afghanistan. There is also a history of these countries working together in Afghanistan against the Taliban movement. India should try and use the side-room politics during the SCO meetings to shape the thinking of these friendly countries in favor of India. Its major interests lie in cooperating in transport infrastructure projects, fighting organized crime and radical Islamic forces, and ensuring its energy security.

• At the same time, India will have to work in Afghanistan with the United States and NATO as well as friendly regional partners. This is important because the United States is likely to be a major player in Afghanistan at least in the short-term, the new U.S. administration's Af-Pak strategy having provided more emphases on engaging Pakistan than India.[56] The Obama administration has committed an aid of some US$1.5 billion a year to Pakistan. As new situations unfold in this region, New Delhi will have to watch these developments carefully and carve out a new engagement strategy with the major players in the region to secure its interests.

Conclusion: How Viable Is Regional Cooperation?

Regional countries seem to have greater stakes in Afghanistan than extra-regional powers. It is imperative that regional countries share the responsibility in Afghanistan and create a favorable atmosphere for dialogue and cooperation to rebuild Afghanistan. The issue of regional cooperation has been echoed in various international and regional forums and conferences on Afghanistan. The UN-backed conference on the future of Afghanistan on March 31, 2009 in the Hague agreed on strengthening security, enhancing regional cooperation, promoting economic growth, and building institutions. Delegates from more than 80 countries and organizations, including Pakistan, Iran, China, Russia, and the Arab world, called for a broader regional approach.[57] The London international conference on Afghanistan was another attempt by the international community to set clear priorities for stabilizing and developing Afghanistan. It acknowledged that "regionally-owned and steered initiatives that showed the need for neighboring and regional partners to work constructively together."[58] In this context, the statement of the recent Istanbul Regional Summit on Friendship and Cooperation is noteworthy. It was pointed out that in 2010 regional partners of Afghanistan will have opportunities to develop and coordinate their activities to advance developmental programs in Afghanistan. The conference welcomed Afghanistan's initiative to invite regional groupings such as SAARC, Regional Economic Cooperation Conference on Afghanistan, SCO, ECO, and OIC to develop as soon as possible a coordinated plan for Afghanistan's regional engagement.[59] In the above context, what is most important to know is how viable regional cooperation is. So far, most of the regional countries have adopted a strategy of engaging Afghanistan bilaterally rather than through a multilateral mechanism. There are various subregional mechanisms at work but not with sufficient results to show.

There is a growing understanding that stability in Afghanistan is linked to the wider region of South, Central, and West Asia, and any strategy to stabilize Afghanistan will require a robust regional economic development plan in addition to more financial aid, troop surge, and good governance for the people of Afghanistan. Peace with the Taliban is neither simple nor easy. In the short term, this is an unachievable task. The success of any strategy to stabilize Afghanistan lies in the capability of the regional and international actors to find new mechanism of cooperation in dealing with the Afghan quagmire. In the current context, the viable option is to pursue a twofold strategy—(1) greater engagement of the regional countries and (2) more pressure on Pakistan to stop supporting the extremist groups in Afghanistan.

Notes

1. Research Fellow, Institute for Defence Studies and Analyses, New Delhi.
2. In the nineteenth century, the Great Game of competition was played between the British and Russian empires and during the cold war, it became an area of competition between the United States and Soviet Union. See J. N. Dixit, *An Afghan Diary*, (Delhi: Konark Publishers, 2000).
3. A majority of the people of Afghanistan belong to Sunni sect of Islam and remaining are Shias. Pashtuns are the majority with separate tribal identities of different subgroups. Other ethnic groups are the Tajiks, Uzbeks and small number of Uigurs and Nuristanis. J. N. Dixit, *An Afghan Diary*, pp. 2–3.

4. Dixit, *An Afghan Diary*, pp. 2–13.

5. In the first seven months of 2009, there has been an average of 898 incidents as compared to 677 during the same time frame in 2008. Incidents involving improvised explosive devices (IEDs) have risen dramatically, to an average of more than eight per day, 60 percent higher than the average during the first seven months of 2008. According to the UN report released on January 13, 2010 2,412 civilians were killed in 2009, a jump of 14 percent over the previous year. See "The Situation in Afghanistan and Its Implications for International Peace and Security," Report of the Secretary General, UN General Assembly Security Council Sixty-Fourth Session, September 22, 2009, A/64/364-S/2009/475, <http://unama.unmissions.org/Portals/UNAMA/SG%20Reports/09sept25-SGreport.pdf> (accessed December 22, 2009); See <http://unama.unmission.org>, (accessed February 16, 2010).

6. On narcotics, he underscored Afghanistan's progress in combating the scourge, stating that "22 out of 34 provinces of Afghanistan were not drug-free," and that last year, "Afghanistan Prosecuted 1,200 traffickers." Nevertheless, he said that an effective fight against narcotics required a "comprehensive strategy and meaningful international cooperation" for addressing "transnational trafficking, precursor drugs and the pattern of consumption trends."

7. "Foreign Minister Sapta" <http://www.mfa.gov.af/detail.asp?Lang=e&Cat=1&ContID=976> (accessed April 12, 2009).

8. <http://www.sectsco.org/html/01651.html> (accessed August 17, 2007); M. Singh Roy, "The Bishkek Summit," <http://www.idsa.in/idsastrategiccomments/TheBishkek Summit_MSRoy_210807> (accessed August 21, 2007).

9. "Afghanistan: Taliban Operating Close to Afghan-Tajik Border," <http://www.eurasianet.org/departments/insight/articles/eav010610a.shtml> (accessed January 8, 2010).

10. "Tajik Minister Says Force Cannot Resolve Afghan Crisis," ITAR-TASS in BBC Monitoring Global Newsline—Central Asia Political, March 28, 2009, <www.monitor.bbc.co.uk> (accessed March 28, 2009).

11. Ibid.

12. <http://www.sectsco.org/html/01651.html> (accessed August 17, 2007); M. Singh Roy, "The Bishkek Summit,"

13. "Kazakhstan to Allocate $50 mln to Train Afghan Specialists," <http://en.rian.ru/world/20091122/156941002.html> (accessed November 24, 2009).

14. "Kazakh Minister Urges Global Cooperation to Solve Afghan Problem," *BBC Monitoring Global Newsline—Central Asia Political*, March 28, 2009 <www.monitor.bbc.co.uk>.

15. M. Aslam Beg, *National Security Diplomacy and Defence* (Foundation for Research on International Environment, National Development and Security, Rawalpindi, 1999), pp. 74–75.

16. Rashid, *Taliban, Islam, Oil and New Great Game in Central Asia* (I.B.Tauris: London, 2000), p. 186; S. S. Pattanaik, "In Pursuit of Strategic Depth: The Changing Dynamics of Pakistan's Afghan Policy," 2006, unpublished paper in M. Singh Roy, "Pakistan's Strategies in Central Asia," Strategic Analysis, vol. 30, no. 4, (2006).

17. M. K. Bhadrakumar, "Afghanistan and the 'Age of Obama'" <http://news.rediff.com/column/2009/dec/01/afghanistan-and-the-age-of-obama.htm> (accessed December 2, 2009).

18. Ibid.

19. "Russia: Lavrov's Full Statement at Opening of Conference on Afghanistan," Ministry of Foreign Affairs, Russia in BBC Monitoring Global Newsline—Former Soviet Union Political, March 28, 2009, <www.monitor.bbc.co.uk>.

20. "SCO Offers to Assist Afghanistan," <http://eng.24.kg/digest/221/> (accessed April 2, 2009).

21. V. Radyuhin, "Russia Returns to Afghanistan, on Request," *The Hindu* (Delhi), June 26, 2008.

22. "Russia Backs NATO Mission in Afghanistan: Lavrov," <http://www.thaindian.com/news-portal/world-news/russia-backs-nato-mission-in-afghanistan-lavrov-lead_100167249.html#ixzz0gQaEh9O4> (accessed January 20, 2010).

23. "Special Envoy of China on Afghanistan Reconstruction," *People's Daily*, January 23, 2006; "China Pledges Nearly $10 Million in Aid to Afghanistan in 2006," The Chinese Government's Official Web Portal, February 1, 2006, <http//www.gov.cn.misc/2006–02/01/content_176548.htm> (accessed February 25, 2007).

24. Ministry of Communications, Islamic Republic of Afghanistan, <http://www.moc.gov.af/vendors.asp> (accessed April 3, 2007).

25. N. Swanstrom, N. Norling, Z. Li, "China," in S. F. Starr (ed.), *The Silk Roads Transport and Trade in Greater Central Asia* (Washington DC: Central Asia-Caucasus Institute Silk Road Studies Program, 2007), p. 403.

26. "India Iran Discuss Terror from Pakistan, Gas Pipeline," <http://www.thaindian.com/newsportal/business/india-iran-discuss-terror-from-pakistan-gas-pipeline_100275525.html>, (accessed November 17, 2009).

27. "Pakistan, Afghanistan, and Russia Hold First Trilateral Summit," <http://www.isria.com/pages/17_june_2009_93.htm> (accessed March 3, 2010).

28. "Pakistan, Afghanistan, Iran Join Hands to Prepare Anti-terror Framework," <http://www.thaindian.com/newsportal/south-asia/pakistan-afghanistan-iran-join-hands-to-prepare-anti-terror-framework_100305256.html> (accessed January 25, 2010).

29. "Saudi Arabia Unable to Solve Taliban Problem in Afghanistan," at BBC Monitoring Global News, South Asia, Political, February 7, 2010.

30. "India Condemns Kabul Attacks," *The Hindu*, (Delhi), February 27, 2010; *Times of India* (Delhi) February 27, 2010.

31. "Pakistan's Balochistan Allegation Is Baseless: Afghanistan," <http://www.thaindian.com/newsportal/uncategorized/pakistans-balochistan-allegation-is-baseless-afghanistan_100223816.html> (accessed January 19, 2010).

32. S. Kapila, "Afghanistan: India Has Legitimate Strategic Interests in Its Stability," <wwww://southasiaanalysis.org/%5Cpapers32%5Cpapers32%5Cpapers3149.html> (accessed February 16, 2010); J. Bajoria, "India-Afghanistan Relations," <http://www.cfr.org/publication/17474/indiaafghanistan_relations.html> (accessed February 16, 2010); S. Ganguly, P. Kapur, "The Unrecognised Benefits of India's Role in Afghanistan," <http://worldpolicy.org/wordpress/2009/05/29/sumit-ganguly-and-paul-kapur-the-unrecognized-benefits-of-india%E2%80%99s-role-in-afghanistan> (accessed March 3, 2010).

33. Ibid.

34. Statement by S. K. Lambah, special envoy of the prime minister of India at <http://indianembassy.ru/cms/index.php?option=com_content&task=view&id=585&Itemid=451> (accessed May 10, 2009).

35. Ibid; K. S. Manjunath, "Living on the Edge, Indian Soldier on for Afghanistan's Reconstruction," April 9, 2009, <http://www.business-standard.com/india/news/livingthe-edge-indians-soldierfor-afghanistan%5Cs-reconstruction/354518/> (accessed March 3, 2010).

36. Statement by Shri S. K. Lambah, special envoy of the prime minister of India, <http://indianembassy.ru/cms/index.php?option=com_content&task=view&id=585&Itemid=451> (accessed May 10, 2009).

37. "India Afghanistan Set Up Partnership Council, to Combat Terror," <http://www.thaindian.com/newsportal/business/india-afghanistan-set-up-partnership-council-to-combat-terror_100224043.html> (accessed August 17, 2009); K. S. Manjunath, "Living on the Edge, Indian Soldier on for Afghanistan's Reconstruction," April 9, 2009, <http://www.business-standard.com/india/news/livingthe-edge-indians-soldierfor-afghanistan%5Cs-reconstruction/354518/> (accessed March 3, 2010).

38. Visit of Foreign Minister Dr. Rangin Dadfar Spanta of Afghanistan, Joint Statement, July 28, 2009 <http://meaindia.nic.in/declarestatement/2009/07/28js01.htm> (accessed November 15, 2009).

39. "India for Continuing International Focus on Afghanistan's Rehabilitation: Krishna," <http://www.thaindian.com/newsportal/india-news/india-for-continuing-internation-al-focus-on-afghanistans-rehabilitation-krishna_100212201.html> (accessed January 20, 2010).

40. "India Afghanistan Set Up Partnership Council, to Combat Terror," <http://www.thaindian.com/newsportal/business/india-afghanistan-set-up-partnership-council-to-combat-terror_100224043.html> (accessed August 17, 2009).

41. E. Schmittand, J. Perlez, "Pakistan Objects to U.S. Plan for Afghan War," *New York Times*, July 21, 2009 cited in S. Simon, J. Stevenson, "Afghanistan: How Much Is Enough?" *Survival*, vol. 51, no. 5 (2009), p. 50.

42. M. Singh Roy, "Pakistan's Strategies in Central Asia," *Strategic Analysis*, vol. 30, no. 4 (2006).

43. M. P. Anil Kumar, "New Delhi Must Hold Its Nerve in the Great Game," February 19, 2010, <http://news.rediff.com/column/2010/feb/19/new-delhi-must-hold-its-nerve-in-the-great-game.htm> (accessed March 3, 2010).

44. R. Karthikeya Gundu, T. C. Schaffer, "India and Pakistan in Afghanistan: Hostile Sports," *South Asia Monitor*, no. 117, Center for Strategic and International Studies, Washington DC, April 3, 2008, <http:// CSIS.org/files/media/CSIS/pubs/sam117.pdf> (accessed April 14, 2009).

45. B. Chellaney, "Surge, Bribe and Run," *The Times of India*, February 3, 2010.

46. R. N. Haass, "What Lies Ahead after the War on Terror?" *The Times of India*, December 24, 2009.

47. Ibid.; S. Chandra, "London Conference on Afghanistan: Implication for India," February 8, 2010, <http://news.rediff.com/column/2010/feb/08/afghan-conference-implications-for-india.htm> (accessed March 3, 2010).

48. M. K. Bhadrakumar, "The Audacity of Afghan Peace Hopes," *The Hindu*, February 4, 2010.

49. "Afghans More Optimistic for Future, Survey Shows," January 11, 2010, <http://news.bbc.co.uk/2/share/bsp/hi/pdfs/11_01_10afghanpoll.pdf > (accessed March 3, 2010).

50. M. K. Bhadrakumar, "Reading the Afghan Equations Correctly," *The Hindu* (Delhi), April 19, 2008, p. 10.

51. C. R. Gharekhan, "Afghanistan—a Way Forward," *The Hindu*, September 24, 2009; Shanti Mariet D'Souza, "Securing India's interests in Afghanistan," *The Hindu*, October 22, 2009.

52. "No Alternative to Dialogue to Resolve India-Pakistan Issue," *The Hindu*, February 28, 2010.

53. M. K. Bhadrakumar, "India Must Energise Regional Diplomacy," *The Hindu* (Delhi), May 15, 2009, p. 8.

54. Ibid.

55. P. Stobdan, "Shanghai Cooperation Organisation and Asian Multilateralism in the Twenty-first Century a Critical Assessment," N. S. Sisodia, V. Krishnappa (eds.), *Global Power Shift and Strategic Transition in Asia* (New Delhi: Academic Foundation, 2009), p. 244.

56. Full text of Obama's Speech on New Strategy in Afghanistan and Pakistan, March 27, 2009.

57. "International Afghan Conference Signals 'New Beginning,'" <http://www.rferl.org/articleprintview/1565188.html> (accessed April 1, 2009).

58. Communiqué of "Afghanistan: The London Conference" Afghan Leadership. Regional Cooperation, International Partnership, <http://afghanistan.hmg.gov.uk/en/conference/communique/> (accessed February 12, 2010).

59. Ibid.

CHAPTER 6

Afghan Factor in Reviving the Sino-Pak Axis

SWARAN SINGH[1]

Situated at the crossroads of the ancient Silk Route network of roads connecting thriving trade as well as other administrative and religious travelers across West, Central, and South Asian societies—and having played
both the buffer and bridge between rising and falling great empires since
ancient times—landlocked Afghanistan today shares its history and borders
with Pakistan, Iran, Tajikistan, Turkmenistan, and Uzbekistan, as also a short
76-kilometer boundary with China. Highlighting its linkages with India,
historians have described it as the gateway for successive invasions into the
Indian subcontinent.[2]

In addition to their interactions since ancient times, the fact that Afghanistan
shares border with Pakistan Occupied Kashmir makes it, at least legally
and potentially, share borders with India. Its induction into South Asian
Association for Regional Cooperation (SAARC) in December 2005 has only
further reinforced closer Indo-Afghan relations. Today, its being a theater and
increasingly a breeding ground for global terrorism has brought it under the
scanner of its neighbors and great powers. In this, both the general situation
in Afghanistan and the increasing Indian presence in Afghanistan and growing bonhomie between the two countries have been an immediate and direct
trigger reinforcing the time-tested India-centric Sino-Pak axis. This in turn
has come to be a major challenge for India's foreign policy though it has yet to
sync with the agenda of India's strategic and foreign policy deliberations.

To begin with, the geostrategic location and difficult rugged terrain—from
snow-covered mountains to barren deserts and rolling steppes—of Afghanistan
has resulted in making it a unique mosaic of multiple ethno-linguistic stocks
with close affinities with all its neighboring societies. However, none of the

great empires of history or dominant and warring domestic communities has ever been able to run its writ in Afghanistan entirely and uncontested. So while Tajiks, Uzbeks, and Turkmens dominate northern Afghanistan, Pushtuns remain in control of much of the south and they have often ruled the country from the capital, Kabul.

Till date, each of these communities remains rooted in its diverse sociocultural moorings and politico-religious affiliations and aspirations and this has been the most perennial challenge for successive Afghan rulers trying to turn this mosaic into a nation-state in tune with modern parlance and paradigms. Otherwise, the people of Afghanistan have had a really long history of habitation in this landmass, with archeological research finding traces of agriculture and pastoral life prevalent here as far back as 10,000 years.[3] Even its current political boundaries can be traced to the end of the nineteenth century and its 1923 Constitution had abrogated slavery and introduced modern social and political institutions and practices, providing for equal rights to all ethnic communities.[4]

In more recent times, the Soviet intervention in Afghanistan from December 1979 and later the rise of radical Taliban and Al Qaeda attracted attention of the international community, yet it is the terrorist attacks of 9/11 and the follow-up U.S. global war on terrorism that have since 2003 brought Afghanistan to the center-stage of the world politics. Especially for its neighbors, although this has since vitiated the whole regional security situation—particularly with threats from terrorism—this has enhanced India's concerns about how this new situation in Afghanistan has reinforced the historic Sino-Pak axis and how, for the first time, the United States is not only looking the other way (as it did in late 1980s) but also openly endorsing and supporting this "all weather" axis of India's two known adversaries. This remains a matter of grave concern in New Delhi and guides its engagement with Afghanistan as also its larger Eurasian policies.

It is in this complex backdrop that this chapter examines the salience of the Sino-Pak axis and how the recent situation in Afghanistan has emerged as a catalyst in facilitating and cementing the time-tested Sino-Pak axis as also its impact on India's threat perceptions and national security initiatives.

Salience of Sino-Pak Axis

The Sino-Pak axis has been far too nuanced and complicated than what has been analyzed in most existing academic and media analysis. No doubt it has been India-centric to a certain extent yet this remains far too broad-based than what has been normally appreciated especially in Indian commentaries that betray their continued vulnerability to political rhetoric and subjective emotion-driven narratives. As a result, these fail to appreciate its multifaceted drivers including the critical role of Afghanistan in reviving and cementing this axis that only further complicates India's perceptions and policies. It is important to appreciate that this axis also has not been an exclusively China-driven enterprise. At the same time, though their common enmity with India remains an important determinant of this axis, its correlation needs to be appreciated with a sense of proportion and perspective. Similarly, the impact of this axis on India's national security interests sure remains immediate and direct yet fluctuating and seasonal.

To frontload our debate on the salience of Afghanistan in reinforcing the Sino-Pak axis, it is worthwhile to deconstruct some of the popular myths about the Sino-Pak axis first. The following nine points allude to the broad crosscurrents that define the real nature of this axis and have kept these two neighbors of India closer together much to India's peril[5]:

(1) The origins of the Sino-Pak axis did not lay in Beijing's one-sided indulgence with Pakistan. Conversely, in January 1950, Pakistan was to become the third non-communist, second commonwealth, and first Muslim country to extend official recognition to the beleaguered and isolated Communist China. That was the period when China needed Pakistan more than the other way round; between 1947 and 1953, Pakistan really showed great magnanimity in warmly reciprocating to China's efforts to break out of its imposed isolation.

(2) Despite India's much celebrated bonhomie with China during early 1950s, Sino-Pak trade totaled US$83.8 million in 1952—almost twice as much as the trade between China and India. It is important to note that this trade was not yet a "defense-centric" trade as it was to become in later years. This situation lasted a few years, yet it underlines how the recent Sino-Indian trade boom is *not* bound to undermine the Sino-Pak axis that remains underwritten by potent politico-strategic drivers.

(3) It was not Pakistan's membership of Western military alliances that attracted China. Instead, for the first five years of its independence, Pakistan had been a vocal supporter of China's membership of the UN. From the end of 1953 though, it had begun to tilt toward the United States,[6] yet the April 1955 Bandung Conference of Afro-Asian nations was to lay the foundations of the Sino-Pak axis.[7] Again, from early 1960s, Pakistan returned to support China's membership of UN, which again was much before the China-India war that is often cited as sole reason for their coming together.[8]

(4) After the first meeting of Prime Ministers Zhou En-lai and Mohammad Ali of Bogra at the Bandung conference, it was not Zhou En-lai but the Pakistani prime minister who visited China first in October 1956. What displayed Pakistani commitment to building ties with China was the fact that Chaudhuri Mohammad Ali (who replaced Mohammad Ali Bogra) had twice

planned to visit China and later his successor H. S. Suhrawardy undertook a 12-day visit to China barely weeks after his becoming prime minister. All this happened in a politically tumultuous period and amidst knowledge of Subrawardy's strong pro-Western leanings.[9]

(5) The Sino–Pak axis have also been understood as guided by China's desire to (a) tie India to South Asia and (b) befriend Muslim neighbors of its Muslim-dominated and turmoil-ridden Xinjiang region that makes Pakistan extremely vital for Beijing.[10] Nevertheless, other than supplying weapons and contributing to infrastructure projects and offering other moral and material support, China has always steered clear of getting entangled in Pakistani military misadventures. This implies that China's current policy of neutrality in Indo-Pak ties is not as anomalous as it is often made out to appear.

(6) China has been seen as exclusively responsible for Pakistan's nuclear weapons and missiles. In reality, the process had begun far earlier with President Eisenhower's "Atoms for Peace" exhibition touring Pakistan in early 1950s. Thoughts of nuclear "weapons" were ignited first in Pakistan by China's nuclear test of 1964 and serious efforts made only after India's nuclear test of 1974 or in the wake of Pakistan's incision in 1971. Even then the first foreign contribution was made by the Netherlands, in the form of smuggling of nuclear blueprints by A. Q. Khan, who was to emerge as the father of Pakistan's nuclear bomb. Also, indirect support of Saudi Arabia, North Korea, Syria, and the United States needs to be acknowledged.[11]

(7) Pakistan has not been the only (or even most favored) destination for China's nuclear and missile transfers. Most clearly established transfer of this nature was China's supply of intermediate range ballistic missiles (IRBMs) to Saudi Arabia in 1988. Experts also insinuate about China's proliferation to other countries such as Libya, North Korea, Iran, Iraq, and Syria.[12] Pakistan may have been the one beneficiary to emerge as a nuclear weapons state, as also the one to further proliferate Chinese technologies and know-how, thus subjecting China to an acid test to withstand international opposition and demonstrate its commitment to the Sino–Pak axis.

(8) Except for certain phases, the Sino–Pak axis has not been confined to defense cooperation. In their initial years, it was their people-to-people contacts that laid the foundations of their equations. Much of their interactions then were between the left-leaning elites of China and the erstwhile East Pakistan. It was only after the incision of Pakistan that West Pakistan and China were to become preoccupied with India's preponderance and with security concerns leading to the Sino–Pak axis becoming defense-centric in nature. Even then, their popular contacts have not been absent and continue to make critical contributions to their mutual goodwill and trust.

(9) Irrespective of Pakistani rhetoric during its wars with India, China has never ever fired even a single shot at India on Pakistan's behest. China's

posturing has always been covert, subtle, and ambiguous.[13] Yet, the Sino-Pak axis has been a constant factor in India's threat assessments resulting in China having "gained public recognition in both Indian and Pakistani assessments as the steadfast ally of Pakistan."[14] However, Indian debates have always underplayed the role of Pakistan in facilitating the August 1971 secret visit to China by Henry Kissinger that was to make both the United States and China obliged to Pakistan.[15]

Their Newfound Motivations

Other than this perceived India-centric nature of the origin of the Sino-Pak axis, Afghanistan has been an important factor in its sustained closer relations for at least the last thirty years. Even the U.S. endorsement of China and Pakistan as frontline states as also of the Sino-Pak axis goes back to the early 1980s when the United States had adopted a strategy of low-intensity conflict to bleed the Soviet forces while in Afghanistan. Accordingly, the first formal endorsement of the Sino-Pak axis by the United States had come as part of the U.S.-Sino-Pak anti-Soviet alliance in raising anti-Soviet Mujahideens in Afghanistan. If anything, the United States was hand-in-glove with the Sino-Pak axis. During the 1980s, the CIA had bought thousands of mules from China, delivered along the Karakoram highway to Pakistan and then to the Mujahideens in Afghanistan. Similarly, most of the weapons used by Mujahideens against the former Soviet Union originated in China.[16] Not since World War II had the Soviets so blatantly expanded their sphere of influence. Although the newfound post-*entente* friends China and the United States intervened on the side of anticommunist forces in Afghanistan, Pakistan became the major supply-line of fighters and provided training grounds that operated from its border regions across the Durand Line.[17]

Given its anti-Soviet nature, this bonhomie was to come to an abrupt end with the withdrawal of Soviet forces from Afghanistan in 1989 followed by the collapse of the Soviet Union. At least for the United States, this time-tested Sino-Pak axis had become not only redundant but also something to be discouraged given its costs in terms of nuclear and missile proliferation. This led to both the United States and China beginning to distance themselves from Pakistan. This was also necessitated by Pakistan's increasingly overt nuclear weapons program as also by its contributions in the creation of the Taliban and then Al Qaeda. This was to lead to Pakistan beginning to become internally vulnerable to these radicalized mercenaries who were suddenly unengaged. Terrorist violence was thus deflected into India's Kashmir with India emerging as its victim as terrorism emerged as a major tool of Pakistan's India policy.[18] This was also to have impact on China with the rising incidence of violence in Xinjiang. Starting from its November 2004 White Paper on National Defence, Beijing had publicly accused Taliban (and Pakistan) of training and equipping Uyghur Muslims spearheading violence in Xinjiang.[19] Pakistan's nuclear tests of 1998 were to make matters worse.

It was the terrorist attacks of 9/11 that were to bring back to Pakistan its status of being the frontline state for both the United States and China, thus reviving the Sino-Pak axis but now in the context of Afghanistan and the global war on terrorism. The inception of the U.S. Af-Pak policy by the Obama presidency in 2009—which shifted the focus of international community away from Iraq's continuing crisis and justified its exponential increase in its military presence in Afghanistan as also the tripling of economic aid to Pakistan—clearly reestablished Pakistan's critical frontline state role in dealing with the proverbial Afghan quagmire. This time, for a change, China has also emerged as a frontline state and its status as the new U.S. interlocutor across Asia goes much beyond its role in Afghanistan. Way back in 2003, following the historic U.S. decision to outsource the nuclear nonproliferation problem of North Korea to China as the convener of the Six Party talks, Washington has been emphasizing on the "stakeholder" status of China in the international system that now seeks to support China's peaceful rise. Result? This emerging centrality of Afghanistan in the U.S. list of priorities has elevated the stature of both China and Pakistan as frontline states, making Afghanistan a unique factor in reviving and cementing the historic Sino-Pak axis, but this time with indirect but clear U.S. support.

What is special this time around is that the United States is not only "looking the other way" but also actively encouraging their bonhomie as it seems least incapable of (even least interested in) enforcing any caution in their mutual indulgence let alone discouraging such reformulations of this axis that have direct and immediate negative influence on India's vital interest and national security. Moreover, rising China and emerging India are also now seen as contenders in presenting parallel paradigms on energy security and resource exploitation as also in sustaining peace on their periphery. Accordingly, as China seeks greater engagement in both Afghanistan with Pakistan, both these countries have come to be an integral part of U.S. Af-Pak strategies. China has since refined its tools of aid, trade, investment, and military cooperation while the United States and Pakistan have been focused on fire-fighting with their militaries fighting terrorism on the ground. An outstretched United States, accordingly, becomes increasingly dependent on China for ensuring efficacy of its Asian policies. China, however, seems plush with both surplus financial resources and manpower that remain its forte.

Beyond these materialistic and militaristic motivations, the circumstances in Afghanistan also offer an opportunity to both China and Pakistan to take another step toward their stated goal of becoming a responsible stakeholder in regional affairs.[20] By contributing to stability in this region, both Beijing and Islamabad seek to allay fears of their likely negative influence. Conversely, both China and Pakistan also need to work together to mitigate the negative influence of other external powers such as India. The rising influence and presence of India in Afghanistan remain a common concern for both Beijing and Islamabad. Speaking at the London Conference on Afghanistan held during January 2010, Chinese foreign minister Yang Jiechi underlined the specific significance of regional initiatives by neighboring countries. He said, "Countries in the region have special associations with Afghanistan due to geographical, religious, ethnic and linguistic reasons."[21] This clearly alluded to Pakistan's unique position as a fellow Islamic republic—one holding Pushtuns, who remain a major ethnic community on both sides of the Durand Line.

India in turn has been conscious of the revival of the Sino-Pak axis and also consistent in making efforts to build bridges both with the United States as with President Karzai's regime. India's pledged support of an unprecedented US$1.3 billion as part of its humanitarian effort in Afghanistan and the unprecedented presence of Indian manpower in various infrastructure-building activities clearly underline India's fast-changing approach to the larger Eurasian reality. Afghanistan is also the first theater where India has deployed paramilitary forces for the protection of its personnel and assets and not as part of a peacekeeping mission. This has been since analyzed threadbare as both the cause as well as the consequence of the Afghan factor emerging as a crucial trigger for the revival of the Sino-Pak axis. At the same time, however, India remains restrained by its wariness in dealing with the so-called moderate Taliban as also with sending its military forces, other than as part of the UN peacekeeping, especially NATO operations in Afghanistan.

China and the Af-Pak Theater

This India logic of the Sino-Pak axis can be traced back to the late 1950s, and China has been Pakistan's main pillar of political support as also main supplier of its conventional weapons as well as nuclear and missile technologies. Bulk of Pakistan's defense production sector is either set up exclusively by China or involves joint projects between these two countries. It was also way back in 1963 that China had promptly resolved its boundary demarcation with both Pakistan and Afghanistan. Although the boundary agreement with Pakistan, demarcating the border between China's Xinjiang and the disputed region of Pakistan Occupied Kashmir, included article VI that keeps this agreement subject to revision depending on the final solution of the Kashmir question, China's agreement on its 76-kilometer border with Afghanistan was full and final. This is also because while for Pakistan China has always been a major neighbor, second only to India, for China during its first thirty years (i.e., between 1949 and 1978), Afghanistan had existed essentially only as a bit player in their relationship with the Soviet Union, India, and Pakistan rather than as an entity that is vitally relevant in its own right.[22]

Until the late 1970s, therefore, Beijing did not perceive Afghanistan as either a matter of military concern or even a religious threat. Having claimed the entire Pamirs in the early 1950s, the Chinese had by the mid-1950s conceded Afghanistan's right to the Wakhan Corridor. Their border negotiations had been noncontroversial and very short—beginning in June 1963 and leading to an agreement on August 2—and culminating in the final treaty signed on November 22, 1963. Early on, the Afghans had also agreed not to allow anti-Chinese activities to be mounted on their soil. This situation was upheld until the April 1978 coup d'état that established the Democratic Republic of Afghanistan. The new regime condemned China for establishing diplomatic relations with the United States and for arming "antirevolutionary" Afghan guerrillas and refugees. Meanwhile, this tumultuous domestic politics in Afghanistan had also been creating problems between Afghanistan and China's close ally Pakistan. By mid-1970s, for instance, situation was agitated enough that Pakistani prime minister Zulfikar Ali Bhutto was talking of a possible war with Afghanistan.[23]

With this rising Chinese consciousness about Afghanistan, the Soviet intervention into Afghanistan from December 1979 was to bring the Sino-Pak axis back into action. Briefly, from the late 1990s, even Pakistan's close links with the Taliban regime in Afghanistan had become an irritant in Sino-Pak equations.[24] But this was not to last too long and 9/11 terrorist attack were to bring them further closer together. The security of terrorism-infected Afghanistan and Pakistan has since become increasingly complicated and intertwined with that of the larger region. This has had a negative influence on the security of neighboring countries including India and China. Rising China sees itself as having both serious stakes and responsibilities in resolving this situation. Moreover, this period has also witnessed a historic evolution in hyphenations, that is, from Indo-Pak ties to Indo-China ties for India; and Indo-Pak ties to Af-Pak theater for Pakistan. This has also reinforced the need for both China and Pakistan to strengthen the Sino-Pak axis. It is common knowledge today that China's relations with Pakistan hold special promise in ensuring the efficacy of its initiatives in Afghanistan and it is with this understanding that Beijing remains seriously engaged with the Karzai regime.

For China, the focus of its collaborations with both Afghanistan and Pakistan also remains part of its larger vision about new threats from energy security, environment, and, above all, terrorism. Therefore, even when

Beijing had a long-term interest in seeing that Pakistan remains a viable counterforce to Indian domination of the Indian subcontinent and the surrounding regions, expanding Sino-Indian economic and political relations have resulted in China counseling Pakistan to enter into discussions with India regarding border disputes and to develop confidence-building measures. Compared to Sino-India trade that stood at US$43 billion in 2009 (and US$52 billion in 2008), Sino-Pak trade though has stayed on the margins and was at US$6.8 billion in 2009, even though they proclaim to raise it to US$15 billion by the end of 2010.[25] In the field of diplomacy though, Pakistan remains the lynchpin of China's South Asia (and even West Asia) policy and both Pakistan and Afghanistan have become increasingly critical of China's looming energy crises and energy security strategies.

In future, therefore, China is likely to return to providing conventional weapons to Pakistan but may cease aiding Pakistan's nuclear program.[26] This is also partly driven by the changing nature of their threat assessments. China's most recent exports to Pakistan, for instance, have been its all-paid counterterrorism mechanisms that include manufacturing Pak Sat-1R and launching it, as also setting up a ground station in Lahore. Similarly, Pakistan has also been expanding its leverages in engaging China. Recent years have witnessed Pakistan projecting its Gwadar Port—which lies 400 kilometers from the Strait of Hormuz in Persian Gulf—as a hub for building a rail, road, and pipeline network to connect Middle Eastern gas and oil to mainland China through the sensitive Xinjiang region. Pakistan has also been at the forefront of leading energy-related initiatives in a SAARC context that makes the Sino-Pak axis critical for China's energy security strategies.

China hopes that, like Pakistan, Afghanistan would become a conduit of rail, roads, and energy pipelines from across the Indian Ocean and Eurasia. Even in the case of Afghanistan, therefore, Beijing remains deeply interested in exploring Afghanistan's untapped gas, oil, iron, copper, gold, uranium, precious gems, and other resources. Terrorism in the region has only reinforced its concerns. Though in all of this China needs Pakistan, Beijing has also been making efforts to indirectly use the Sino-Pak axis in directly engaging Afghanistan. For instance, even before 9/11, having failed to persuade (or enable) Pakistan to rein in the negative influence of its Mujahideens and the Taliban into its turmoil-ridden Muslim majority autonomous region of Xinjiang, China was one of the few countries that had successfully established links with the ruling pariah Taliban regime.[27] China was also wary of how the United States had used 9/11 and its actions against the notorious Taliban regime to establish its military presence across Central Asia, a wariness that had been the other driving force for the Sino-Pak axis. This made Beijing use its influence in the Shanghai Cooperation Organisation (SCO), invite Pakistan and Afghanistan as observers into the SCO, all to restrain and if possible remove U.S. military presence in the Central Asian Republics (CARs).

Having been ousted from Central Asia, the United States have been seeking expanded engagement with South Asian countries as an alternate route to the CARs. This U.S. engagement with Pakistan (as also India and Afghanistan) has also reinforced China's need for reviving its special relations with Pakistan. So, in addition to its time-tested Karakoram Highway, China has been planning on rail and road links between Urumqi (Xinjiang) and Hawailian, as also on contributing to the Pakistani plan to build new roads into Afghanistan including those linking Torkham with Khandahar and Chaman with Jalalabad. Similarly, as President Karzai opened Afghanistan to foreign investments in 2007, China's Metallurgical Group won rights to develop the world's largest copper fields in Aynak for US$3.5 billion. Likewise it has also invested in other infrastructure-building projects such as hospitals, irrigation and restoration projects, and laying (over 200,000) phone lines across Afghanistan.

Amongst the other drivers of the revived Sino-Pak axis in Afghanistan, both China and Pakistan have been extremely conscious of India's unprecedented presence and goodwill in Afghanistan. This is viewed by both China and Pakistan with concern though it is Pakistan that has been far more vocal and paranoid and lately it has been raising this bogey about Indian involvement in promoting violence in its northwestern borders, especially in Baluchistan. Commentators have linked increasing terrorist attack on Indian personnel in Afghanistan to Pakistan's expanding its combat zone beyond Kashmir. Similarly, given the tumultuous nature of Sino-Indian relations in the last few years, China has felt tempted to take the hard-line stance. To quote from the *China Daily* editorial of February 2010:

> It is clear that without Pakistan's cooperation, the US cannot win the war on terror. Therefore, to safeguard its own interests in the fight against terrorism in South Asia, the US must ensure a stable domestic and international environment for Pakistan and ease the tension between Pakistan and India. This makes it easy to understand why Obama appointed Richard Holbrooke as special envoy for Afghanistan and Pakistan issues, and why India is included in Holbrooke's first foreign visit. In fact, the "Afghan problem", the "Pakistani problem" and the "Indian-Pakistani problem" are all related.

For Indian commentators, given that Beijing is fully aware of India's distaste for third-party intervention in Kashmir, especially if the third-party is Beijing or recommended by Beijing, such comments are a direct suggestion that Beijing is talking of what Pakistan calls the "core issue"—that is, Kashmir.[28]

India's Limitations and Leverages

No doubt Pakistan remains China's most valued ally in the region, yet China and India have also found themselves thinking alike on so many vital issues that space seems to have opened for India policymakers to deal with the growing influence of the Sino-Pak axis in this region. Compared to the United States and Pakistan, both India and China remain wary of "moderate" Taliban formulations. China has even been blaming U.S. forces for aggravating political and social turmoil, while asking it to end its military presence in Afghanistan and recommending UN-supervised international peacekeeping as the most viable alternative. India is in favor of this argument though New Delhi has its own normative reasons to be concerned with U.S. military presence in Pakistan and Afghanistan. However, India is uncomfortable with the idea of privileging an intra-Afghanistan settlement inclusive of "moderate" Taliban, an idea that both China and Pakistan (also the United States) feel inclined to support. It is interesting to note that China has never blamed Pakistan for propping up the Taliban and Al Qaeda, whose activity in the region remains a major reason for the presence of NATO and U.S. forces.

From the perspective of its concerns about increasing NATO and U.S. forces in its periphery, China is too happy to bolster Pakistani military in dealing with both Al Qaeda and the Taliban on both sides of the Durand Line. China's own intervention in Afghanistan has been distinctly confined to infrastructure development and resource exploitation. China remains ever ready to provide manpower for development projects in Afghanistan to bolster alternative strategies for its relief and reconstruction. There are reports that in view of NATO and U.S. forces not being able to stabilize the situation, China may send even ground troops—a contribution that could bring great change in China's role in this region. But that, of course, is not in sync with Beijing's currently favored strategy that hopes to continue engaging the United States and the EU to ensure that they continue to take military action in Afghanistan and, at the same time, to maintain modest pressure for their early exit. China has to deftly maximize benefits and minimize its costs, which include costs from the continued presence of U.S. and NATO troops in Afghanistan. And on this its interests clearly converge with those of Pakistan, making Afghanistan an important factor in reinforcing the Sino-Pak axis.

U.S. engagement of both China and Pakistan in their Af-Pak policy provides an opportunity for China to project itself as a responsible regional power and partner in contributing to the stability and development of a troubled region in its neighborhood. And in case the United States and NATO succeed in banishing terrorism from Afghanistan this will only enhance China's geopolitical position and links with Pakistan, provided Beijing is able to keep the pressure and ensure that the United States and NATO exit Afghanistan as soon as possible. That explains why China is using economic instruments that will help it stay put hoping that military instruments will have to leave as soon as violence in subdued and, if this comes true, then for China partnership with Pakistan in tackling Afghanistan could not present a better opportunity for expanding its influence in the region in future. Having learned from the United States' and India's increasing presence in Afghanistan, both China and Pakistan remain determined to evolve mutual understanding and shared visions and to finesse their joint strategies and expand their cooperation to ensure the marginalization and exit of their common adversaries from their immediate backyard.

Notes

1. Professor of Diplomacy and Disarmament, School of International Studies, Jawaharlal Nehru University, New Delhi. Author can be reached at ssingh@mail.jnu.ac.in

2. M. Ewans, *Conflict in Afghanistan: Studies in Asymmetric Warfare* (New York: Routledge, 2005), p. 19.

3. M. Ewans, *Afghanistan: A New History* (London: Routledge Curzon, 2002), p. 10.

4. H. Emadi, *Culture and Customs of Afghanistan* (Westport, CT: Greenwood Press, 2005), p. 31; A. Basu Ray, "Modernisation in Afghanistan (1880–1901): Futile Experiments?" in M. Singh (ed.), *Asia Annual 2001* (New Delhi: Shipra Publications, 2001), pp. 139–141.

5. For detailed debate see S. Singh, "Introduction," in S. Singh (ed.), *China-Pakistan Strategic Cooperation: Indian Perspectives* (New Delhi: Manohar, 2007).

6. W. J. Barnds, *India, Pakistan and the Great Powers* (New York: Praeger, 1972), pp. 95, 323.

7. A. H. Syed, *China and Pakistan: Diplomacy of an Entente Cordiale* (Karachi: Oxford University Press, 1974), p. 54.

8. The first visible change in China-Pakistan relations was to come as part of the July 1961 U.S. visit by President Ayub Khan declaring to vote in favor of seating China at the UN and UNSC. Later, the fact that the United States had extended military help to India during the China-India War was to further make Islamabad skeptical of U.S. intentions and grow closer to China, which also now had strong reasons to befriend India's archrival.

9. A. Syed, *China and Pakistan: Diplomacy of an Entente Cordiale* (London: Oxford University Press, 1974), p. 140; M. Ahmed, *Government and Politics in Pakistan* (Karachi: Pakistan Publishing House, 1963), p. 68; S. Noor Ahmad, *Marshal La Se Marshal La* (Lahore: Deen Mohammadi Press, 1965), p. 480; Shorish Kashmiri, *Husain Shaheed Suhrawardy* (Lahore: Chatan Publications, 1967).

10. M. Ehsan Ahrari, "China, Pakistan and the 'Taliban Syndrome,'" *Asian Survey*, vol. 40 no. 4 (2000), pp. 666–668.

11. R. Einhorn, "China and Non-proliferation," *In the National Interest*, vol. 2, no. 13 (2003), <http://www.inthenationalinterest.com/Articles/Vol2Issue13/Vol2Issue13einhorn. html> (accessed September 2010).

12. D. Shen, "China and Nuclear Proliferation: General Policy with Application to South Asia," in M. Kumar Kayathwal (ed.), *Security and Foreign Policy in South Asia* (Jaipur: Pointer Publishers, 1999), p. 82; S. A. Kan, "Chinese Missile and Nuclear Proliferation: Issues for Congress," *CRS Issue Brief*, IB92056, June 12, 1992, p. CRS-4.

13. F. Grare, *Pakistan: In the Face of the Afghan Conflict 1979–1985 at the Turn of the Cold War* (New Delhi: India Research Press, 2003), p. 32.

14. S. Bhutani, *A Clash of Political Cultures: Sino-Indian Relations, 1952–1962* (New Delhi: Roli Books, 2004), p. 207; General Mohammad Musa (Retd.), *My Version: India-Pakistan War 1965* (Lahore: Wajidalis Ltd., 1983), pp. 11, 92.

15. B. R. Deepak, *India and China 1904–2004: A Century of Peace and Conflict* (New Delhi: Manak, 2005), p. 465.

16. Y. Shichor, "The Great Wall of Steel: Military and Strategy in Xinjiang," in S. F. Starr (ed.), *Xinjang: China's Muslim borderland* (New York: M. E. Sharpe, 2004), p. 145.

17. J. F. Kornberg, J. R. Faust, *China in World Politics: Policies, Processes, Prospects* (Boulder, CO: Lynne Rienner, 2005), p. 175.

18. B. M. Chengappa, *Pakistan, Islamisation, Army and Foreign Policy*, (New Delhi: APH Publishing Corporation, 2004), p. 127.

19. S. Singh, "China's Afghan Policy: Limitations versus Leverages," in K. Warikoo (ed.), *The Afghan Crisis: Issues and Perspectives* (New Delhi: Bhavana Books, 2002), pp. 411–414.

20. K. Slaten, "China's Bigger Role in Pakistan, Afghanistan," *South China Morning Post* (Hong Kong), February 12, 2009, <http://www.carnegieendowment.org/publications/index.cfm?fa=view&id=22735> (accessed February 26, 2010).

21. "China Supports Trilateral Cooperation among Turkey, Afghanistan, Pakistan," website of the Embassy of the People's Republic of China in the Islamic Republic of Pakistan, dated January 27, 2010, <http://pk.chineseembassy.org/eng/zbgx/t653913.htm> (accessed February 26, 2010).

22. S. Dutta, "China and the Afghan Conflict," in V. D. Chopra (ed.), *Afghanistan: Geneva Accord and After* (New Delhi: Patriot Publishers, 1988), p. 120.

23. H. St. Amant Bradsher, *Afghanistan and the Soviet Union* (Durham, NC: Duke University Press, 1983), pp. 11, 63.

24. Y. Shichor, "The Great Wall of Steel," p. 144.

25. S. Singh, "Limitations of China-India Economic Engagement," *China Report* (New Delhi), vol. 44, no. 4 (2010), also "China Set to Become 4th Biggest Trade Partner of Pakistan: MOFCOM Official," <http://pk.chineseembassy.org/eng/zbgx/t657322.htm> (accessed March 1, 2010).

26. J. F. Kornberg, J. R. Faust, *China in World Politics: Policies, Processes, Prospects*, p. 177.

27. M. Ehsan Ehrari, "China, Pakistan and the 'Taliban Syndrome,'" *Asian Survey* (California), vol. XL, no. 4 (2000), p. 658; S. Gangadharan, "The China-Taliban Equations," *Akrosh* (New Delhi), vol. 6, no. 6 (2000), p. 62.

28. M. K. Bhadrakumar, "China Breaks Its Silence on Afghanistan," *Asia Times* (Hong Kong), February 25, 2010, <http://www.atimes.com/atimes/China/KB25Ad03.html> (accessed February 26, 2010).

PART II

India and China in Central Asia, between Cooperation, Parallelism, and Competition

India and China in Central Asia: Mirroring Their Bilateral Relations

Jean-François Huchet[1]

This chapter analyses the extent of the Sino-Indian diplomatic thaw since the early 1990s. Without ignoring the existence of multiple cooperation channels, or seeking to minimize the importance of the considerable achievements realized in recent years by the two governments toward normalizing their relations, the attempt here is to show that relations between the two Asian giants remain hamstrung by a series of geostrategic and economic rivalries. Despite fast growth in trade and in specific areas of economic cooperation, the normalization of ties between Beijing and New Delhi does not yet constitute a genuine strategic partnership. Central Asia, as a new "hunting ground" for energy resources and geopolitical influence for regional and global players, will be no exception to the current nature of the bilateral relationship: competition and sometimes conflict (though not on an open basis) will probably dominate their relationship in this region, and cooperation will be established only for pragmatic reasons.

After three decades of "freeze" following the war between the two countries in November 1962, India and China resumed diplomatic and trade exchanges in the early 1990s. Since the beginning of the new century, bilateral diplomatic relations have improved dramatically, with each year seeing several meetings between heads of government/state as well as ministers.[2] The contrast is arresting when one compares the period of tension and freeze that characterized Sino-Indian ties from the late 1950s. Over the past decade, there has been a multidirectional acceleration of official contacts. Although it is still too early to assess the real impact of this diplomatic warming, it is nevertheless interesting, given the growing weight of the two Asian giants on the world arena, to consider the nature and evolution of their relationship in the context of international relations and global trade. Have the two countries really buried the hatchet and overcome the mutual suspicion that stemmed from the 1962 military conflict? Are they embarking on a fraternal entente of the kind symbolized by the early 1950s slogan "Hindi-Chini Bhai-Bhai"[3] that was so dear to Nehru? More generally, could these warming diplomatic and trade relations give rise to a strategic partnership similar to that between the United States and Britain, or between post–World War II Germany and France, leading to the emergence of an Asian regional integrating force and a major axis in international relations?

This chapter counsels caution with regard to media hyperbole as well as misleading conclusions drawn from similarities between the two countries' population size or pace of economic growth. Without ignoring the existence of several channels of cooperation or seeking to minimize the importance of considerable achievements realized in recent years by the two governments toward normalizing their relations, this chapter argues that pragmatism will prevail on both sides of the Himalayas. The chapter concludes with a consideration of the impact of the current state of this relationship on the two countries' expansion in Central Asia.

Geostrategic Suspicions and Rivalries in Bilateral Relations

The war that China unleashed on October 20, 1962 lasted a mere 30 days, but it continues to haunt Beijing–New Delhi relations nearly half a century later. Quite apart from the territorial differences over recognizing the McMahon line,[4] and the question of which side was responsible for sparking the war,[5] geopolitical tensions of a more general nature have persisted since the 1950s, despite transformations in the regional and international contexts. Moreover, the alliance between Pakistan and China as well as India's backing of the Dalai Lama continue to weigh heavily on Sino-Indian ties.

The Clash of Two Nationalisms and the Border Conflict

The border conflict continues to poison bilateral relations. Many a skirmish occurred after the 1962 ceasefire. The two armies confronted each other twice, in 1967 (in Sikkim) and in 1984 (in the Sumdorong Chu Valley in the state of Arunachal Pradesh in northeastern India). In 1987, the tone rose again in both capitals, sparking fears of another conflict. It was only in 1993 that New Delhi and Beijing signed an agreement on preserving "peace and tranquility" along the line of control. Despite setting up a working group in November 2003 to resolve the border conflict, 13 rounds of negotiations have yet to yield concrete results, and the occurrence of many recent incidents along the line of control shows that a resolution of the border issue is still far away. The Chinese government still voices complaints when senior Indian leaders visit Arunachal Pradesh considering that this disputed region is part of Tibet and thus part of China.[6] Indian leaders in Arunachal Pradesh have on many occasions conveyed to New Delhi their fears over the reinforcement of Chinese military deployments since 2005. Indian authorities have officially complained of hundreds of incursions by Chinese troops across the line of control since 2006.[7]

Nearly 65 incursions by the Chinese forces have been recorded since the start of 2008 in Sikkim, even though Beijing ostensibly acknowledged New Delhi's sovereignty over the region following a visit to China in 2003 by India's prime minister at the time, Atal Bihari Vajpayee. New Chinese demands over Sikkim could serve as a means to gain the upper hand in negotiations with India over another disputed region, that of Tawang in Arunachal Pradesh. Apart from possessing major mineral resources and being situated in the strategic Brahmaputra valley, Tawang is a noted centre of Tibetan Buddhism. The Galden Namgey Lhatse Monastery, where the sixth Dalai Lama was born, is the second largest in Tibetan Buddhism after the Potala in Lhasa, and members of the Tibetan government in exile made it an important base after escaping from Chinese control.[8] It is, therefore, understandable that the Chinese communist leadership wants to recover the territory (the Tibetan issue is further discussed below).

It is worth noting that the border issue has become a kind of barometer of bilateral ties. Despite the thaw in relations, New Delhi and Beijing are still blowing hot and cold over the settlement of the border dispute, a function of evolution in the most important bilateral issues. The Indo-U.S. rapprochement, especially on the civilian nuclear issue, has led to a series of statements on China's part with regard to the border dispute. The most astounding—though it merely reiterated a known Chinese stand—was that of the then Chinese ambassador Sun Yuxi, shortly after President Hu Jintao's India trip. The envoy, in November 2006 in a televised interview, said that Arunachal Pradesh belonged to China, drawing a predictably furious response from India.[9] On the Chinese side the border dispute is largely a geostrategic security issue of little concern to the public in comparison with its conflicts with Japan and Taiwan; on the Indian side, however, the 1962 humiliation left a great deal of nationalism and emotion invested. Indian defense officials have recently reiterated that their forces lag far behind China's in terms of strength.[10] In case of failure to push back a strong Chinese offensive, the Indian side will have only limited elbow room in negotiations.

As for rivalry over Asian leadership, economic and diplomatic clout has turned the wheel largely in Beijing's favor since the late 1950s. China has a permanent seat in the UN Security Council, and its rapid economic development has led to active diplomacy with Asian countries since the mid-1990s. A recent article by Tarique Niazi points out that China has built a network of control over the Indian sphere of influence in Asia through its tentacles in ASEAN, the Shanghai Cooperation Organisation (SCO), and the South Asian Association for Regional Cooperation (SAARC),[11] and also through its highly active bilateral engagement with India's neighbors, including Nepal, Bangladesh, Myanmar, and, of course, Pakistan.[12] Moreover, China has developed naval cooperation with many Asian countries to form what the Indian media calls a "pearl necklace" in South and Southeast Asia through the establishment of a series of permanent military bases to secure energy supplies.[13] Some of these bases—such as those in Chittagong in Bangladesh, Coco Islands in Myanmar, Habantota in Sri Lanka, Marao in the Maldives, and Gwadar in Pakistan—are very much in India's maritime "zone of influence."

Faced with all these actions on China's part, India has responded by obtaining dialogue partner status with ASEAN and observer status in the SCO, and by launching some regional initiatives such as the Mekong Ganga Cooperation[14] in November 2000. Despite this Indian counteroffensive, China clearly dominates the diplomatic game of spheres of influence. India hasn't renounced its ambitions for Asian leadership, however. Nearly a half century after Nehru's death, and in an Asia transformed in the economic and political spheres, India continues, rightly or wrongly, to believe that its demographic weight, democratic values, and more recent economic dynamism could help it play a major role on the Asian stage. This hasn't gone unnoticed in Beijing, and the Sino-Indian rivalry in Asia continues to fuel mutual suspicion comparable to that between France and Prussia in Europe in the latter half of the nineteenth century. For now, Beijing enjoys a clear advantage.

The last issue that is symptomatic of this geostrategic rivalry concerns India's aspirations for a permanent seat on the UN Security Council. Beijing has blown hot and cold over the issue, never officially setting out a clear position. On May 30, 2008, during a BRIC (Brazil, Russia, India, China) summit at Yekaterinburg in Russia, China refused to sign a Russian-drafted final communiqué backing India's candidature. This refusal has been interpreted to mean either a change in China's position or a pursuit of a policy of control over the Indian sphere of influence. Such handling of an issue so dear to India shows a deep chasm in bilateral strategic cooperation.

Fears of Containment and "Ménage à Trois" with the United States

Among the geopolitical tensions that bedeviled the two countries during the 1950s, China's fear of being encircled through India's strategic military alliance with the United States and the Soviet Union contributed greatly to the deterioration of bilateral ties. President Bill Clinton's visit to India in 2000, coming 22 years after Jimmy Carter's in 1978, revived containment fears in the Chinese leadership, which believed that the United States was seeking to choke off China's emerging economic might through alliances with Japan, South Korea, and the countries of Central and Southern Asia. These fears deepened when a proposal for a nuclear accord was unveiled during Prime Minister Manmohan Singh's official U.S. trip in July 2005 and ratified by both countries during the fall of 2008. The proposal envisaged India (a non-signatory to the Nuclear Nonproliferation Treaty—NPT) allowing inspectors of the International Atomic Energy Agency (IAEA) access to its nuclear energy production sites,[15] along with a moratorium on nuclear tests. In exchange, India gained access to U.S. nuclear technology, ending 34 years of embargo on trade in nuclear material following India's first test in 1974.[16] More generally, the agreement allows India to officially enter the select group of nuclear powers while remaining outside of the NPT.

More generally and beyond the nuclear issue, there have been extensive ideological departures in both the United States and India over the last few years on the question of bilateral ties. Washington has decided to pay serious attention to New Delhi's political, military, and economic weight in the region and to integrate it in strategic action plans for Asian security.[17] On the Indian side, both the Bharatiya Janata Party (BJP) and the Congress have worked for rapprochement with the United States since early this decade.[18] This change has been reflected in a clear increase in bilateral cooperation in the strategic, military (joint exercises in the Indian Ocean), and economic domains.

While on the Chinese side the effort is to limit the effects of strategic and military encirclement, on the Indian side the rapprochement with the United States is aimed at directly limiting Chinese influence in Asia. One analyst of India's foreign policy put it this way:

> More than the sops offered by the United States, it is the stick being wielded by China that is powering the current Indian sprint toward a substantive strategic partnership with the United States, one that can only affect China's interests in Asia negatively. The Himalayan chill now enveloping Sino-Indian[19] ties is creating summer warmth in India's relations with the other superpower, the United States.[20]

This profound and lasting change in Indo–U.S. ties has not escaped the attention of Beijing, which fiercely attacked the nuclear accord.[21] Beijing will henceforth be in a more uncomfortable position in countering the strategic repercussions of the Indo–U.S. rapprochement. Beijing could also decide to cooperate earnestly with India so that the United States does not become India's only and privileged partner in nuclear issues. Engaging in nuclear cooperation with India would also help Beijing test New Delhi's nonaligned credentials. And being "courted" by both Washington and Beijing on the issue, New Delhi might be tempted to play them off against each other for higher stakes. However, keeping equidistant ties with Washington and Beijing and pursuing nonalignment would be a difficult balancing act for New Delhi. The United States would need some proof of exclusive loyalty from India in order to subdue Washington skeptics. Any perception in Washington that New Delhi is trying to play the big powers against each other could recoil and discredit India (as in the past) in its role of new Asian strategic partner.

This "ménage à trois," which is clearly transforming strategic relations in Asia in a deep and abiding way, could have surprises in store. Beijing holds many advantages—close ties with Pakistan, weight in the UN, special relations with the ASEAN, and economic might in the region—to counter India's influence. But as the nuclear issue has clearly shown, even after a 30-year hiatus, China's fear of containment through warming Indo-U.S. ties is again becoming a crucial factor in Sino-Indian ties. This development can only exacerbate feelings of suspicion and pragmatism on both sides.

The Tibet Issue

The Dalai Lama had considered requesting political asylum as early as in 1956, when he was part of the official Chinese delegation during Premier Zhou Enlai's visit to India. Sino-Indian relations were then in fine fettle. In an agreement signed with China in April 1954, India had officially recognized that Tibet belonged to China, and Nehru had signaled to the Dalai Lama that he did not wish to interfere in Sino-Tibetan affairs for fear of annoying the Chinese government. But by the time the Dalai Lama crossed the border in March 1959 and sought refuge in India, Sino-Indian ties had already begun unraveling. The arrival of the Dalai Lama in India in late March 1959 further vitiated Sino-Indian ties. In an interview to the American journalist Edgar Snow in October 1960, Zhou Enlai said the boundary dispute "came to the fore" after "the Dalai Lama had run away." He accused India of wanting to "turn China's Tibet region into a 'buffer zone.'" He said, "They don't want Tibet to become a Socialist Tibet, as had other places in China," and that "the Indian side...is using the Sino-Indian boundary question as a card against progressive forces at home and as capital for obtaining 'foreign aid.'"[22]

Nearly a half century after the Dalai Lama's arrival in India, the Tibet issue continues to bedevil bilateral ties. The Sino-Indian modus operandi over Tibet remains fragile. Although the Indian authorities periodically pull up the Dalai Lama when he uses his Dharamsala headquarters for political activities to which China objects, he enjoys free movement in and out of India and manages to irritate the Chinese on a regular basis.[23] The major uprising in March 2008 by the Tibetan population again upset whatever understanding existed between China and India on the Tibet issue.[24] Despite the arrest of some Tibetans for staging "anti-China" activities on Indian soil,[25] the Indian authorities have allowed demonstrations in many major cities, and the foreign ministry shed its reticence for once in calling for negotiations between the Dalai Lama and the Chinese authorities as well as for a nonviolent resolution of the troubles in Tibet.[26]

Yet another matter of Sino-Indian dispute and one linked to Tibet and the border issue is Beijing's claim to the whole of Arunachal Pradesh. China argues that the state was part of Tibet before the signing of the McMahon Treaty in 1914. Once Tibet became Chinese territory, Beijing believed Arunachal Pradesh belonged to it by right, insisting that Tibet had signed the McMahon Treaty under British military pressure. From the late 1950s, Beijing has claimed that the treaty stemmed from colonialism and was thus worthless, and that India, which had suffered under British domination, ought not to recognize it. During the Sino-Tibetan talks in July 2007, the Arunachal Pradesh issue was on the agenda for the first time. China voiced firm opposition to the decision taken by the Tibetan government in exile in December 2006 to accept India's sovereignty over Arunachal Pradesh, including the region of Tawang that hosts the Galden Namgey Lhatse Monastery (see section above on border conflict). Beijing's accusation that the Tibetan government in exile had succumbed to pressure from Indian leaders, public opinion, and media was rejected in both Dharamsala and New Delhi.

The Sino-Pak Alliance

The warming of Sino-Indian diplomatic relations since the early 1990s has not dented Sino-Pak ties. If anything, during this period an economic facet has been added to their strong military and strategic ties. Notable was the launch in 2002 of development of a deep-water port at Gwadar, close to the Strait of Hormuz, through which nearly 20 percent of global oil moves. The construction of the Gwadar Port, which is now operational, has been 80 percent funded by China (a total of US$250 million) and overseen mainly by state-owned China Harbour Engineering Co. Ltd., with nearly 350 Chinese engineers engaged in the project.[27] In 2007, the Pakistani government entrusted operation of the port to the PSA Group of Singapore for 25 years and conferred duty-free status to Gwadar for 40 years. Although the management agreement does not give Chinese ships exclusive rights to the use of the Gwadar Port, given its assistance in the construction and the solidity of Sino-Pak cooperation, China can put to good use the strategic site to protect its energy supplies and boost its military presence. The project, emblematic of the strengthening of naval cooperation between Pakistan and China, has serious strategic and military implications for India, according to W. Lawrence S. Prabhakar. India will face a more muscular Chinese naval presence and a greater Chinese effort to stifle New Delhi's influence in the Indian Ocean region.[28]

More generally, although Sino-Pak relations have had to gradually adapt to the thaw in Sino-Indian ties, Pakistan remains a highly strategic card in China's foreign policy "game" in the region. A realignment with India or even a balancing act favoring New Delhi appears highly unlikely. Beijing may well deny it, but the fact is that Pakistan remains a cat's paw limiting India's moves on the regional chessboard: the links between Pakistan's intelligence agencies and Islamic militants impeding a resolution of the Kashmir dispute (among other problems), military rivalry with India, communal conflicts between Hindus and Muslims that poison Indian polity, the paralysis in the SAARC caused by the Indo-Pak rivalry, and the access China has gained to Pakistan through the Strait of Hormuz, all constitute destabilizing factors for India and will restrain its influence in the region. Moreover, China also enjoys a privileged entry into the Muslim world via Pakistan. Finally, the Indo-Pak rivalry allows China to keep its military presence relatively limited on its southwest flank. Despite transformations in its relations with the United States and Russia, China must still ensure the security of its borders (at 22,722 kilometers, it is the world's longest, exceeding even Russia's) with its 14 neighbors, including the 4,057-kilometer stretch with India (China's third longest border after that with either Mongolia or Russia). Moreover, China's Taiwan policy and the U.S. military presence in Japan and South Korea preclude the concentration of too much military might in one border area.

For all these reasons, Sino-Pak relations can be expected to take precedence over Sino-Indian ties. As for New Delhi, although more and more voices call for a reasonable and rational view of Sino-Pak ties, the fact remains that the relationship is largely seen as a strategic alliance shrouded in secrecy and mainly aimed against India.

Beyond Complementarities in Economic Relations

A Window-Dressing Pact between the "World's Workshop" and the "World's Back Office"

Among the analyses touting the potential for strategic economic cooperation between China and India, the alliance between the "world's back office" (India) and the "world's workshop" (China) is certainly one that has captured the imagination of many people. Nevertheless the question arises as to how this specialization in two different segments of the information industry, and more generally between two economic sectors (manufacturing in China and services in India), could induce greater bilateral cooperation. Few factors favor a deeper industrial cooperation between the two sides beyond superficial complementarities.

First, cooperation would require each side to renounce over the long term the development of the segment in which it holds the lesser advantage, both absolute and comparative (in the sense David Ricardo meant). The current industrial policies of the two governments as well as the decisions of companies in each country show a trend in exactly the opposite direction. Over the past few years, the two sides have, with varying degrees of success, striven to correct the economic development trajectories that led to an atrophy of industry in India's case and a comparatively slower growth in services in China.

More generally, China's domestic market is as large as India's. Unlike in countries with small populations that need to specialize, vast domestic markets favor the development of the broadest range of industries and services with help from foreign direct investment. A country's comparative advantage and competitiveness in particular industrial sectors are not fixed in time; they can change and improve gradually over the years behind an evolving and selective protectionist wall. Like France, Germany, and the United States in the nineteenth century,[29] or Japan, Korea, and Taiwan, China followed this policy in the 1980s and 1990s before agreeing to significantly reduce tariffs in order to gain entry into the World Trade Organization (WTO) in 2001. A recent study showed that in India too, many industrial sectors have gained considerable international comparative advantage in just a few years.[30] It should be stated here that if the day dawns when China and India dominate manufacturing or services, it will almost certainly take place in the context of competition, and not in the framework of a complementary partnership.

Second, even if current complementarities in economic activities (or trade) persist, they will not necessarily lead to closer industrial or technological cooperation between the two countries. The example of Japan and China shows this quite clearly. Despite rapid increases in trade—Japan is China's second largest trade partner after the EU—and the existence of complementarities in economic activities, the two countries have failed to develop a strategic partnership in the technological or industrial sectors. Indeed, such partnerships are an exception rather than a rule in contemporary international economic relations in general. Technological and industrial cooperation among the EU countries was preceded and accompanied by a process of political integration that is unique in economic history today. It is also worth mentioning the great access enjoyed by Japanese and then Korean and Taiwanese firms to U.S. technology following World War II. This cooperation was largely motivated by geopolitical considerations linked to the defense of American interests and the development of a "capitalist front" during the cold war. Sino-Indian ties fit neither the political integration model nor that of military protection or domination. It is, therefore, highly likely that current complementarities in economic structures will not lead to closer technological and industrial cooperation, but rather to a more "conventional" progress in trade and investment exchanges, which in itself would be a great improvement over the past.

Thus the presence of firms such as China's Huawei in Bangalore or Tata Consulting in Pudong[31] must be viewed with prudence and sobriety. China has become a formidable market globally. The growth in demand for information technology services provided by Chinese firms as well as the presence of multinationals in China are such that it would be suicidal for Indian companies that lead in this domain not to boost their presence in China, all the more so as a number of multinationals that have been long-term clients of Indian firms have gained a major presence in China and continue to require servicing by Indian companies in the framework of traditional software subcontracting. In India, Chinese companies have sought to increase their presence since early this decade as part of a process of rapid multinationalization. This is as imperative for them as it is for their Indian, Western, or Japanese counterparts. They have to strengthen and diversify their acquisition of technological competence or risk losing out not only to global competition but also within a domestic market that has opened under WTO membership. But in no way does it signal the development of closer bilateral strategic technological and industrial cooperation as some analysts suggest.[32]

Current Complementarities and Future Competition in Trade Relations

As graph 7.1 shows, bilateral trade, which was almost nonexistent until the 1990s (US$260 million in 1991), has grown rapidly since the start of the new century. The targets set by the two countries in 2005, of achieving US$20 billion worth of trade by 2008 and US$30 billion by 2010, have been surpassed: bilateral trade was already worth nearly US$51.2 billion in 2008 (US$43 billion in 2009).

Research by M.S. Qureshi and Wan Guanghua[33] into the foreign trade structures of both countries shows that the rapid rise in bilateral exchanges can be explained by increasing complementarities between India and China in recent years. India exports mainly raw materials to China, which for its part ships mostly manufactured goods to India. Rapid economic growth in both countries and the "reserves of complementarities" in products traded increase the likelihood of rapid bilateral exchanges in the years to come, the authors say. India, for instance, has strong potential for increasing exports of leather and inorganic chemicals to China.[34] At the same time, China is capable of boosting its exports of telecommunication products and computers to India. Similarly, there is potential for intersectoral growth in steel, organic and inorganic chemicals, as well as machinery.[35] With a growth rate of 50 percent since 2000, bilateral trade could total US$100 billion by 2015.[36]

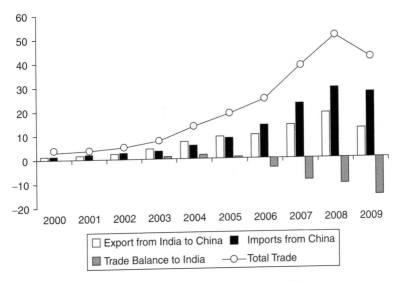

Graph 7.1   Evolution of India-China Bilateral Trade (in billion US$)

There are, nevertheless, many potential roadblocks to this expansion in bilateral trade, some of which could ignite trade disputes. The first is linked to the nature of Indian exports to China, mostly agricultural raw materials and minerals. Supplies of these natural resources are not unlimited, and given its domestic needs, India could be forced to limit the export of some products to China. This already occurred in July 2008 when Chinese importers, who had greatly boosted their imports of Indian iron ore, were faced with a new 15 percent tax on export of the raw material, which the government in New Delhi imposed in order to ensure adequate supplies to domestic steelworks. It would not be surprising if such protectionist measures multiply in India, given the enormous demand for mineral resources in China and the corresponding exponential demand from India's own industry.

The second limit is linked to the imbalance in trade that favors China, portending disputes. After having registered a balanced trade and even a favorable turn between 2003 and 2005, India's deficit vis-à-vis China has risen rapidly since 2006 to reach US$16 billion in 2009 (see graph 7.1), or 17 percent of India's total trade deficit.[37] With a nearly 10 percent share of India's total imports in 2009, China is India's second trade partner, just behind the United States in 2008 (US$32.44 billion of imports from the United States and total trade amounts to US$63.63 billion in 2008). Despite a rapid rise in exports to China, India accounts for a mere 2 percent of China's total imports. Projections by foreign trade experts on the effects of a Sino-Indian FTA show that China would benefit more. A 2007 study by the Institute for Economic Research in Munich showed that an FTA with a 100 percent reduction in tariffs would result in China's exports (with 2004 as base year) rising by 131 percent, against a mere 38 percent in India's case.[38]

In short, the rapid rise in exchanges is an encouraging and positive sign of warming Sino-Indian relations. It also ends the aberrant situation of what until early in this decade was a ridiculously tiny trade volume, given the rate of economic growth in the two countries, their physical proximity, and the structural complementarities of their economies. However, as the example of Sino-Japanese ties shows, growing trade exchanges do not equate to strategic economic cooperation. Moreover, the sharp rise in trade could potentially bring an imbalance for India that could lead to disputes with China. At the conclusion of the latest Joint Economic Group meeting of the two countries in Beijing in January 2010, the Indian commerce ministry handed to its counterpart a list of specific items (fruits, meat, software, TV programs) on which India expects China to take urgent action.[39] Therefore, India is most likely to continue protecting its industry and raw materials vis-à-vis China, which for its part is unlikely to be content with its current strong position in trade and will keep pushing for an opening of India's domestic market to its goods. Tensions and conflicts are thus likely to accompany soaring bilateral trade.

Parallel and Competitive Quests for New Sources of Raw Material

In January 2006, when Mani Shankar Aiyar, then minister for petroleum and natural gas, signed an agreement during an official trip to China, many analysts and commentators enthusiastically endorsed an energy alliance between the two sides.[40] The agreement envisages extensive cooperation in the fields of oil and gas prospecting, refining, and distribution, as well as promotion of alternative energy. Above all, the two countries expressed their willingness to eschew cutthroat competition in the search for energy in all corners of the globe in order to meet their soaring demand for energy supplies. Just before the agreement was signed, in November 2005, India's ONGC (Oil and Natural Gas Corporation) and its Chinese counterpart, CNPC (China National Petroleum Corporation), got together for the first time ever to buy a major stake in Petro-Canada's venture in Syria's al-Furat oilfields.[41] The January 2006 tie-up gave a genuine boost to Sino-Indian energy cooperation. Many other large-scale joint operations have since been launched, such as those in Sudan[42] and Colombia,[43] and more recently in the supply of equipment for electricity generation.[44] The idea of two economic giants and new entrants on the international energy chessboard getting together to take on major U.S., European, and Japanese companies on their favorite hunting grounds of Africa, the Middle East, Latin America, and Central Asia was no doubt a seductive one for many observers. Nevertheless, as in other areas of bilateral exchanges, the extent of Sino-Indian cooperation in the field of energy seems to have been greatly exaggerated.

First of all, an analysis of the behavior of Indian and Chinese firms shows that the relationship is more typically characterized by often brutal competition. Since 2004, Indian firms in direct competition with Chinese companies have lost a series of bids for oil projects in Angola, Kazakhstan, Ecuador, and Myanmar.[45] In a much rarer instance of an Indian firm upstaging a Chinese firm with much greater experience in this field, ONGC Videsh Limited (OVL, the foreign operations arm of ONGC) outbid SINOPEC (China Petroleum and Chemical Corporation) to acquire Imperial Energy, a British firm that controlled parts of oil projects in Russia and Kazakhstan, for US$2.6 billion.[46] How could competition be avoided when securing foreign energy supplies has become a key national priority for both giants? The International Energy Agency estimates that India's dependence on foreign energy sources will rise from 73 percent (in 2006) to 91.6 percent by 2020. In China, which was energy self-sufficient until the mid-1990s, dependence on foreign sources is expected to rise to 50 percent in 2010 and to 77 percent in 2030.[47] Thus the two countries' national independence relies in part on access to new foreign energy sources.

Sino-Indian competition is not limited to hydrocarbons; it is equally fierce in the quest for other raw materials both countries lack. Phosphates, iron ore, nickel, lead, scrap iron, aluminium, leather, zinc, and tungsten (despite major domestic resources of these in China's case) are the subject of intense competition, not only between China and India, but also among other big countries.[48]

Finally, even when large Indian and Chinese energy groups decide to collaborate, there is nothing to indicate that these projects would be conducted in the framework of a strategic and special relationship at the expense of their cooperation with other countries. On the contrary, an analysis of the foreign cooperative ties of the two countries' oil and gas giants shows that their links to European, U.S., Russian, Japanese, Saudi, and Australian companies and to major developing countries such as Brazil, Chile, Iran, and Venezuela are much longer, more numerous, and sustained. As of late 2006, China National Offshore Oil Corporation (CNOOC) had signed 182 contracts and agreements in the oil domain with 76 companies from 21 countries.[49] Shell, Exxon, Chevron, and Total are CNOOC's major partners, far ahead of any Indian companies. Of CNOOC's 50 oilfields offshore of China, 27 are exploited jointly with foreign companies, but no Indian firm was among them as of late 2006. Similarly, there is no indication that Chinese firms are about to overtake the number of cooperative ventures Indian companies have established with Western and Japanese companies or with those from other developing countries. ONGC Videsh, for instance, has much closer links to Brazil's Petrobras, having signed a series of strategic agreements in 2007.[50] Agreements signed with CNPC or SINOPEC are in no way exclusive or different from many others that ONGC Videsh has signed with foreign partners since early this decade.[51] The same holds true for the Indian private group Reliance.[52] Sino-Indian cooperation in energy is thus devoid of any exclusivity; the importance of recently launched joint projects is put in perspective by the older and more sustained ties both sides maintain with other countries. In recent years China and India have treated each other as potential alliance partners for specific projects in the same way as they have other state and private actors on the world energy scene.

Given these facts, it is clear that despite the implementation of a few joint projects in hydrocarbons since 2005, the two countries are far from forging a strategic relationship in energy and raw materials. Cooperation between Chinese and Indian firms appears for now to be more the exception than the rule as the two engage in ruthless competition to secure new foreign energy supplies.

Pragmatic Alliances in International Economic Institutions

Aware of the quasi-hegemonic role of the United States and Europe in the Bretton Woods institutions created after the end of World War II, Jawaharlal Nehru had, after India's independence in 1947, actively sought Chinese support to counter the weight of the developed countries and to defend the interests of developing ones. The Bandung conference in April 1955, with 29 developing countries represented, was a culmination of this strategy, in which India played a central role. The impossibility of the Chinese Communist regime's participation in the Bretton Woods institutions until the 1970s (when China at last took the seat occupied by Taiwan), as well as the 1962 Sino-Indian war, put paid to any hopes of building a strategic Beijing–New Delhi axis aimed at defending the interests of developing countries. Although China was reintegrated gradually into all international organizations during the 1970s (UN) and 1980s (IMF, World Bank), it was above all with the WTO accession in 2001, at a time of far-reaching reforms to the world trade architecture, that the opportunity for a Beijing–New Delhi axis arose again. The Doha Development Round of the WTO, which commenced in 2001, was marked by strong differences between the developed and developing countries, with India playing a particularly active role in defending the latter's interests.

As a newcomer in the WTO, China kept a low profile in the Doha Round. Nevertheless, Beijing clearly aligned itself with New Delhi's stand, letting India play the leader's role during ministerial conferences. China and India have also defended shared stands in international negotiations on climate change. The two countries have made common cause over the past several years, affirming this entente forcefully at the Climate Change Conference in Bali in December 2007[53] and more recently in Copenhagen in December 2009 by rejecting all attempts by developed countries (including the United States) to impose quotas on the emission of greenhouse gases.[54]

Seeing the impact of such alliances between India and China, several analysts have spoken of a new era in international economic organizations, especially the WTO. Unthinkable only 15 years ago, given the state of their bilateral ties and China's limited participation in international economic organizations, such alliances could be repeated in future and could change the current power equations in international negotiations. Nevertheless, such Sino-Indian alliances can be expected to be limited to specific issues and remain pragmatic and they are unlikely to lead to the formation of a systematic axis for defending developing countries. The numerous geostrategic rivalries examined in the first part of this chapter limit the extent of such alliances, which are still largely dictated by specific convergences linked to the defense of national interests and limited to some economic issues. The lack of alliance between the two countries on other matters, such as the reform of the United Nations and the IMF, was demonstrated by China's attitude toward India's aspirations to obtain a permanent seat in the UN Security Council.[55] Even in a regional institution such as the Asian Development Bank, China avidly guards its interests: Last year, Beijing vetoed an ADB loan to India for infrastructure development in Arunachal Pradesh.

Conclusion

In analyzing various facets of Sino-Indian relations, the effort here has been to show that behind the media and official hyperbole over warming diplomatic and economic ties since the early 1990s, the bilateral process remains entangled in a series of geostrategic and economic rivalries. Beijing-New Delhi relations continue to be dogged by mutual suspicion inherited from the 1962 war, itself the result of a clash of mutually irreconcilable nationalisms and ambitions for dominance in the Asian scene. In 1959—that is, three years before the start of the war—Jawaharlal Nehru had told Edgar Snow in an interview that "the basic reason for the Sino-Indian dispute was that they were both 'new nations' in that both were newly independent and under dynamic nationalistic leaderships, and in a sense were 'meeting' at their frontiers for the first time in history."[56] The Indian historian Ramachandra Guha takes this further, insisting that the India-China conflict "was a clash of national myths, national egos, national insecurities and—ultimately and inevitably—national armies."[57]

The geopolitical context has, of course, changed greatly today, but the clash of two nationalisms on the Asian scene continues to influence Sino-Indian relations.

A relationship based on pragmatism is the best that can emerge from this loaded and complex heritage and its three decades of "freeze." As the above analysis indicates, Beijing-New Delhi ties today are subject more to tensions that pit them against each other than joint projects that bring them closer together. Normalization of relations does not, therefore, imply the emergence of a genuine strategic partnership; the hypothesis according to which development of bilateral exchanges would help sweep away rivalries, tensions, and mutual suspicions appears to lack credibility. The "gentle commerce" between people and nations, so dear to Montesquieu and Adam Smith, today helps China and India to (re)discover each other,[58] and to act together on some specific issues of energy and environment and in the WTO framework. But that does not amount to a magic wand that will wish away the numerous problems currently weighing down bilateral relations as well as both countries' hegemonic ambitions in Asia. In other words, India and China are well on their way to changing the world.[59] But they do so separately, through the emergence of their respective economic and military might, rather than through a strategic partnership, which for the moment remains more a pious declaration than a reality.

As far as Indian and Chinese expansion in Central Asia is concerned, there are few possibilities for them to forge a strategic partnership to cooperate. What has been stated in this chapter on a more general level about the bilateral relationship will also characterize their interaction in this part of the world. Central Asia is considered by both India and China as a strategic region for their geopolitical activities and energy procurements and transportation, the very areas where competition dominates their relationship.

As other chapters in this volume show, China is far ahead of India in north of Central Asia in terms of political influence, trade volume, and energy procurements. To counter Russian—and more marginally U.S. and European—influence, one could imagine good reasons for China to forge an alliance with India. But so far, as examples of oil and uranium procurements show, China, conscious of its strength, thinks that it can go it alone, and there are no signs indicating that it would be ready to alter this strategy. Therefore, China has no incentives to cooperate with India in this part of Central Asia. Quite the contrary, with India becoming more assertive and dynamic in Central Asia, China perceives India as a new competitor. Cooperation could be possible, but again based on pragmatic reasons and for a limited objective.

As for the southern and western part of Central Asia, China does not want to see India recover its historical (including the one existing during the British period) sphere of influence. This part of Central Asia (especially Pakistan, Kashmir, and Afghanistan, and to a lesser extent Tajikistan and Turkmenistan) is critical for both countries. First, as other chapters in this book point out, oil and gas transportation from Iran and the Middle East to India would be much easier if peace and economic stability would prevail in this region. It would also give the edge to India over China on energy transportation from Iran and the Middle East. Given the fact that India already has a natural advantage on China in the Indian Ocean on energy maritime route, China would probably be worried to see India gaining such a privileged position on energy transportation. Second, as we have shown in this chapter, Pakistan, and indirectly Kashmir, remains crucial for China to check India's sphere of influence in Central Asia. All in all, there are again more reasons to see realpolitik dominating China's diplomatic attitude vis-à-vis India concerning this part of Central Asia. This is not to say that China would incite further destabilization of an already much troubled region. But as far as Indian influence in this region is concerned, which is again a critical point for Beijing's diplomacy, status quo is preferable.

Notes

1. Director of the Centre d'Etudes Français sur la Chine Contemporaine (French Centre for Research on Contemporary China—CEFC), Hong Kong
2. The last bilateral visit was Prime Minister Manmohan Singh's to Beijing in January 2008, which followed President Hu Jintao's trip to India in November 2006.
3. The slogan was launched during Zhou's visit in 1956 to signify brotherly ties between China and India.
4. The demarcation line set in the Shimla Accord signed in 1914 with Tibet—without Chinese assent—by British foreign secretary for India, Henry McMahon.
5. See N. Maxwell, *India's China War* (London: Pantheon Books, 1970), on New Delhi's responsibility in the conflict; see also J. W. Garver, *Why Did China Go to War with India in 1962?* <http://www.people.fas.harvard.edu/~johnston/garver.pdf> (accessed March 2, 2010); and J. B. Calvin, *The China—India Border War (1962)* (Marine Corps Command and Staff College, April 1984), <http://www.globalsecurity.org/military/library/report/1984/CJB.htm> (accessed March 2, 2010).
6. China strongly criticized the visit of Manmohan Singh to Arunachal Pradesh on October 3, 2009. The Indian prime minister was campaigning on behalf of his ruling Congress Party for the local legislative assembly elections, *Financial Times*, October 19, 2009.
7. The Indian government protested some 140 intrusions beyond the LoC by the Chinese army in 2007. *Times of India*, April 6, 2008. The number of incursions was reported to have doubled in 2008. *Christian Science Monitor*, September 29, 2009.
8. The Dalai Lama made a four-day visit to Tawang on November 8, 2009. It came 50 years after his stay in the region following his escape from Tibet in 1959. The 2009 visit provoked a strong reaction from the Chinese government, see *The Irish Times*, November 10, 2009.
9. For an analysis of the political repercussions in India from this statement, see *The Hindu*, November 24, 2006.
10. Air Chief Marshal P. V. Naik was reported to have said in September 2009 that the Indian aircraft strength was a third of China's and that India was trying to augment its strength. *Christian Science Monitor*, September 29, 2009; Army Lieutenant General B. S. Jaswal was reported to have said in January 2010 that India lagged "far behind" China in infrastructure development along the border. *Hindustan Times*, January 23, 2010.
11. In which China has won observer status.

12. T. Niazi, "Sino-Indian Rivalry for Pan-Asian Leadership," *China Brief*, vol. 6, no. 4, February 15, 2006.

13. O. Zajec, "La Chine affirme ses ambitions navales," *Le Monde Diplomatique*, September 2008, no. 654, pp. 18–19.

14. With six member countries—Cambodia, India, Laos, Myanmar, Thailand and Vietnam—to promote regional cooperation in education, tourism, transport and culture.

15. The agreement provides for placing 14 of India's 22 reactors under IAEA supervision.

16. For details of the agreement, see Jayshree Bajoria, "The U.S.-India Nuclear Deal," *Council on Foreign Relations*, February 2008, <http://www.cfr.org/publication/9663/usindia_nuclear_deal.html?breadcrumb=%2Fbios%2Fbio%3Fgroupby%3D0%26hide%3D1%26id%3D13611%26page%3D3> (accessed March 1, 2010).

17. R. N. Burns, "America's Strategic Opportunity with India: The New U.S-India Partnership," *Foreign Affairs*, vol. 86, no. 6 (2007), pp. 131–146.

18. E. Luce, *In Spite of the Gods: The Strange Rise of Modern India* (New York: Random House, 2007), chapter 7, pp. 261–299.

19. This followed the statement in January 2007 by China's ambassador in India, Sun Yuxi, who claimed that the entire state of Arunachal Pradesh belonged to China.

20. M. D. Nalapat, "Himalayan Chill over Sino-Indian Relations," UPI Asia.com, July 30, 2007, <http://www.upiasia.com/Politics/2007/07/30/commentary_himalayan_chill_over_sinoindian_relations/1008/> (accessed March 1, 2010).

21. For China's official reaction to the Indo-U.S. agreement, see M. Malik, "China Responds to the U.S.-India Nuclear Deal," *China Brief*, vol. 6, no. 7, March 29, 2006.

22. In *Look Magazine*, October 18, 1960, quoted by E. Snow, *The Other Side of the River: Red China Today* (New York: Random House, 1963), and cited by Ramachandra Guha, *India after Gandhi: The History of the World's Largest Democracy* (London: Picador India, 2007), p. 319.

23. During Zhou's last pre-war visit in April 1960, finance minister Morarji Desai (who went on to become prime minister in 1977 after the end of the state of emergency declared by Indira Gandhi) told him: "Lenin sought asylum in the United Kingdom but nobody restricted his political activities." In Guha, *India after Gandhi*, p. 318.

24. See R. Barnett, "Thunder from Tibet," *New York Review of Books*, vol. 55, no. 9, May 29, 2008.

25. "India Detains Tibet Protestors," *Time Magazine*, March 14, 2008.

26. "India Breaks Silence, Calls for Talks on Tibet," *The Times of India*, March 16, 2008.

27. *Associated Press,* May 13, 2007.

28. W. Lawrence, S. Prabhakar, "The Maritime Convergence," in Swaran Singh (ed.), *China-Pakistan Strategic Cooperation: Indian Perspectives* (New Delhi: Manohar & Centre de Sciences Humaines, 2007), pp. 231–247; see also O. Zajec, "La Chine affirme ses ambitions navales."

29. J. Frieden, *Global Capitalism: Its Fall and Rise in the Twentieth Century* (New York: W. W. Norton & Company, 2006).

30. A. Batra, Z. Khan, *Revealed Comparative Advantage: An Analysis for India and China* (New Delhi: ICRIER working paper, no. 168, August 2005).

31. Tata Consulting opened its first China office in 2004 in Shanghai's Pudong technology development zone.

32. Gillian Goh Hui Lynn, *China and India: Towards Greater Cooperation and Exchange* (working paper, Lee Kuan Yew School of Public Policy, National University of Singapore, 2006).

33. M. S. Qureshi, G. Wan, *Trade Expansion of China and India: Threat or Opportunity* (United Nations University, UNU-WIDER, research paper, no. 2008/08).

34. T. P. Bhat, Atulan Guha, Mahua Paul, and Partha Pratim Sahu, *India and China: Building Complementarities and Competitiveness into WTO Regime* (New Delhi: Bookwell Publishers, 2008).

35. M. S. Qureshi, G. Wan., *Trade Expansion of China and India: Threat or Opportunity.*

36. T. P. Bhat, *India-China Free Trade Area. Is It Feasible?* (New Delhi: ISID working paper, 2008), p. 20.

37. India's trade deficit stood at US$76 billion for the April–December 2009 period, or for the entire year at US$91 billion, see Government of India, Ministry of Commerce and Industry, Department of Commerce, Economic Division, 01.02.2010, <http://commerce.nic.in/tradestats/indiatrade_press.asp>.

38. S. K. Bhattacharya, B. N. Bhattacharyay, *Gains and Losses of India-China Trade Cooperation—A Gravity Model Impact Analysis* (Munich: CESifo working paper, no. 1970, April 2007), p. 18.

39. *The Indian Express*, January 21, 2010.

40. See *Asia Times*, Internet edition, September 26, 2006, <http://www.atimes.com/atimes/South_Asia/HI26Df01.html> (accessed March 1, 2010).

41. G. Sachdeva, "India's Attitude towards China's Growing Influence in Central Asia," *China and Eurasia Forum Quarterly*, vol. 4, no. 3 (2006), pp. 23–34.

42. The two sides worked together in extracting, transporting, and refining oil in Sudan. After Western and Japanese firms pulled out over the Darfur conflict, China, Malaysia, and India took charge of oil-related works in the country.

43. CNPC and ONGC, though subsidiaries, invested US$850 million in a 50 percent stake in Omimex Colombia (U.S. subsidiary of Omimex Resources) with oil and gas extraction in South America.

44. In August 2008, the Indian Reliance group, along with Shanghai Electric, signed a five-year industrial cooperation agreement envisaging design, production, and marketing of electricity generation equipment; "India-China Sign over US$1bn Energy Pact," <http://www.2point6billion.com/2008/08/18/india-china-us3bn-energy-pact/> (accessed March 1, 2010).

45. In August 2005, China National Petroleum Corporation paid US$4.18 billion to acquire Canadian-owned PetroKazakhstan, outbidding India's Oil and Natural Gas Corporation, which offered US$3.9 billion. See J. Nandakumar, "India-China Energy Cooperation: Attaining New Heights," *Institute for Defence Studies and Analyses Strategic Comments*, November 28, 2005; and on Angola, AFP, October 15, 2004.

46. *International Business Time*, August 28, 2008.

47. *World Energy Outlook 2006* (Paris: International Energy Agency, 2007), p. 101.

48. In *Ecomine*, various issues, Direction Générale de l'énergie et des matières premières, Ministère de l'Industrie (Directorate general of energy and raw materials, French industry ministry) <http://www.industrie.gouv.fr/energie/publi/f1e_pub.htm> (accessed March 2, 2010).

49. <http://www.cnooc.com.cn/yyww/hzsy/dwhz/244281.shtml> (accessed March 2, 2010).

50. See <http://www.ongcvidesh.com/display1.asp?fol_name=News&file_name=news135&get_pic=ovl_news&p_title=&curr_f=135&tot_file=155> (accessed March 2, 2010).

51. Notably with those from Brazil, Italy, Russia and Venezuela.

52. Report of activities, 2003, 2004, 2005, 2006, 2007.

53. *The Hindu*, December 18, 2007.

54. See C. Paskal, S. Savitt, "Copenhagen Consequences for the U.S., China and India," *UPI Asia.com*, January 4, 2010, <http://www.upiasia.com/Politics/2010/01/04/copenhagen_consequences_for_the_us_china_and_india/1537/> (accessed March 2, 2010).

55. *Washington Post*, January 31, 2010.

56. E. Snow, *The Other Side of the River: Red China Today*, p. 761.

57. Guha, *India after Gandhi*, p. 336.

58. A. Sen, "Passage to China," *The New York Review of Books*, vol. 51, no. 19, December 2, 2004.

59. T. Khanna, *Billions of Entrepreneurs: How China and India Are Reshaping Their Futures—and Yours* (Boston: Harvard Business School Press, 2007).

CHAPTER 8

India-China Interactions in Central Asia through the Prism of Paul Kennedy's Analysis of Great Powers

BASUDEB CHAUDHURI AND MANPREET SETHI[1]

Paul Kennedy in his treatise *Rise and Fall of Great Powers*[2] identified the attributes of a Great Power. In his study of five centuries, he discovered three features that particularly stood out in the case of a nation that already was or likely to be a Great Power: military prowess to support the nation's strategic position; economic conditions to ensure a consistent supply of wealth from a flourishing productive base (at home or abroad) buoyed by superior technology to support the requisite military power; capacity to act independently of other great powers or without their support in the pursuit of objectives deemed critical for national interest.

Written in late 1980s in the context of world players of that time, the attributes that the book identified for great power status have not significantly changed. However, in 20 years, the great powers or the candidates for great power status are indeed no longer the same. Although the United States is still preponderant, albeit in relative decline, Russia and EU are far less important than what Kennedy had conceived. The Soviet Union lost territory, resources, and diplomatic clout as a result of its break up in the end of 1991. The European Union has emerged as a powerful economic bloc but without enough political influence. Meanwhile, China (which Kennedy did see as a future great power) and India (which found no mention in the book) are clearly powers to be reckoned with in the future.

This chapter performs two tasks. First, it takes a look at whether China and India could qualify as the new great powers in the sense that Kennedy had defined. Are they acquiring the necessary attributes in terms of economic strength, military potential, and assertiveness in their foreign policy? Second, it examines the interactions of these two nations in Central Asia. Every great power has had a geographical space from which it could draw material resources for fortifying its economic and military prowess, in which it could flex its power or showcase its influence. Would China and India use Central Asia in this capacity? How will their political, economic, and energy engagements with Central Asian nations pan out? And, will the two rising powers be able to cooperate or find themselves in competition, or even rivalry, with one another in the region?

Do China and India Qualify as Great Powers?

A number of articles and books in recent times describe China as the new "Great Power" or "Rising Power" and India as an "Emerging Power." The tendency to compare the two—in terms of size, economic and military capability, future potential, successes and failures, and their rise from a situation of great poverty and backwardness in the middle of the twentieth century—is normal. In fact, books have also been written (for example, Hochraich's 2007 work *Pourquoi l'Inde et La Chine ne domineront pas le Monde de demain*, to mention only one) explaining why China and India will *not* be the dominant powers of the decades to come. On a number of conventional indicators—such as per capita income, technological innovation and its relationship to military capability, the growth of factor productivity, the share of world trade, and voice in world affairs—the positions of China and India are still far from those of the United States in the middle of the twentieth century, even though China is well ahead of India (see table 8.1 for some overall comparisons). So does it make any sense to speak of the Great Power status of these two countries at this point of time?

More than absolute figures, it is through a newly demonstrated capacity to transform and initiate change, especially economic change in a very short period of time, by the standards of world history, that these two countries have captured world attention. Let us not forget that in the 1960s the Club of Rome had predicted that both countries would be decimated by poverty and famines. They had been written off as failed nations and, in fact, as failed systems, both economically and politically. The surprising element is their capability to provide strong economic competition in the developed nations' traditional areas of dominance such as industry, services, and high-end human resources. Equally surprising is their decision to adopt what is turning out to be an aggressive model of capitalism: commerce and entrepreneurship driven with overt and covert State support, and with a willingness to use protection to develop indigenous industries and then opening up gradually to the rest

Table 8.1   China-India Global Comparison

	China	India
GDP per capita in PPP	5,963	2,762
Pop below $1/day	9.9 %	34.3%
Adult literacy	91%	61%
Urban population	40%	29%
Life expectancy (years)	73	64
Infant mortality per 1000 live births	23	56
Doctors per 100,000 people	106	60
Mobile phones/100 persons	53	38
Internet users/100 persons	22	7
Road density (km per 100 sq km)	21	114
CO_2 emissions (metric tons per person)	4	1
Share of world trade	7.8%	1.5–2%

Source: Times of India, Special Issue, October 3, 2009

of the world—a time-tested formula of the developed nations. It is also note-worthy that of all the major developed economies that have been struggling to stay afloat amidst the global financial crisis that broke out in 2009, the only two economies that have managed to register a positive growth, though lower than what they had experienced in the past few years, have been China and India.

This is hardly a small achievement given that the countries suffered a process of systematic de-industrialization during its two centuries of effective colonization since mid-eighteenth century. During this period, the total share of China and India in the global manufacturing output fell to abysmal levels, from a peak time when it accounted for 58 percent of the world's manufactured goods while Great Britain produced only 3 percent. However, by the mid-1940s when the new political and economic world order took shape, the economic reality stood almost reversed. It has taken China and India a good part of the twentieth century to find their feet and reemerge to respectable positions of economic growth. China, of course, is way ahead of India, having embarked on its economic reforms more than a decade before India did so in the early 1990s. However, the footprint of their rapid economic development is spreading across the developed and developing worlds.

The reforms that were carried out in both countries were driven by "ruling coalitions" of elites in two very different political systems, but which share the common characteristic of building support, if not a semblance of broad consensus, through complex negotiating and institutional mechanisms. The tacit support of political bases—the peasantry and the industrial middle class in China, and the middle and "subaltern" classes in India who started perceiving what Hirschmann called the "tunnel" effect in development[3]—gave reforming elites the necessary legitimacy for the implementation of reforms. In the Indian context, the electorate systematically voted in favor of reform and change, never hesitating to throw out incumbent governments that did not perform and deliver tangible gains.

Today, in both countries, a new political, intellectual, and industrial elite is trying to get rid of a major element of the intellectual hangover or the intellectual handicap left behind by colonial domination that influenced the Nehru-Mao years—the blanket rejection of the market economy as a force for economic change, because of the perceived link between rise of the market economy with imperialism and colonialism. The civilizational depth—the long history of trade, commerce, industry, statecraft, learning, and the place of knowledge in society—seems to be playing a positive role in both countries.

At the same time, there is another factor that has not been taken into account sufficiently. This is the psychology of the ruling elite and the middle classes of both nations who show a desire for a "*revanche sur l'histoire*," a willingness to contest on their own terrain the great Western powers to acquire what they consider to be their rightful place in history. No amount of statistics can take into account the role of collective psychology in nation building, and the ruling elites of both nations have marshaled this resource with considerable dexterity in a globalizing world. The media and public opinion in both countries are actively pushing their governments to adopt a hard-line stance on many major issues of potential conflict, irrespective of the perceived stature of the adversary. In such public stances, there is a fair element of what Chomsky called "the fabrication of consent" by ruling elites.

Evidently then, China and India of today manifest many of the attributes identified by Kennedy. Although it is true that their economic strength, technological prowess, or military might is nowhere close to that of the other predominant nations, the trend is quite evident. China has over a trillion dollars in its foreign exchange reserves, it boasts of the highest economic growth rate, and its GDP is widely projected to take over that of the United States by 2030. Increased financial resources enable greater military spending and hence the country's armed forces are undergoing drastic modernization, including in the nuclear dimension. China's official pronouncements, particularly on national defense, have been published with far greater regularity over the last decade. In fact, its White Papers on National Defence (WPND) express a greater confidence in articulating Chinese national objectives, including that of "reunification." These documents also reflect how China has gradually shed its shyness on the role it envisages for its military in achieving national objectives. The 2004 White Paper, for instance, boldly identified the priority of the PLA as building "capability for winning both command of the sea and command of the air and conducting strategic counter-strike."[4] WPND 2006 outlined a three-step modernization strategy for Chinese armed forces: "The first step is to lay a solid foundation by 2010, the second is to make major progress around 2020 and the third is to basically reach the strategic goal of building informationized armed forces and being capable of winning wars by the mid 21st century."[5] The ASAT test conducted by China in January 2007 announced the country's ambitions to use space for its military activities. The PLAAF commander has more recently expressed China's ambitions to "develop and deploy offensive and defensive weapons in space" as a "Great Wall of steel in the blue sky."[6]

Meanwhile, China has also significantly enhanced its political legitimacy by becoming an active participant in the nonproliferation regime (it plays an active role in resolving the North Korean nuclear crisis). It has asserted its regional clout through creation of institutions such as the Shanghai Cooperation Organization (SCO) and its role in ASEAN. It has also built its goodwill with smaller neighbors, settled its border disputes with all, except India, and is providing ample economic assistance to all.

India, however, is an emerging and aspiring power. Its economic reforms were hesitantly initiated in 1991. Its sustained economic growth since then, its nuclear muscle flexing and military modernization, as well as its ability to manage its own crises—military conflict (the Kargil War, in 1999), or humanitarian (the tsunami, in 2004)—have brought forth the potential of the country as a major player in the region and beyond. India's huge market potential for defense and strategic goods and technology makes it an attractive destination for many of the developed countries. Also, its command in the information technology (IT) sector and its strength as an English language user add to its attraction as a business proposition. Also, India's nuclear rapprochement with the United States in 2008 and the growing engagement between the two nations in the fields of defense, military, and other strategic spheres have provided India with a new stature in international relations.

Both China and India have shown several times since the 1950s the capacity to act militarily to defend their strategic interests. They have also shown a willingness to stand apart, even isolated and under sanctions, for an act that was considered necessary for national security.[7] The degree of success in these interventions has not been uniform and there have been notable failures, but the willingness to act has been there. In more recent times, however, there is a greater tendency to replace direct military intervention (such as that of the Chinese in Korea in the 1950s and of the Indians in Sri Lanka in the 1980s) with use of a combination of diplomacy and the threat of military or naval power. Is this behavior based on the fact that they are not yet at the level of military strength acquired by the United States or the Soviet Union (including

Russian intervention in Chechnya) in the recent past? Or, is it because India and China have quickly (more quickly than the United States or Russia) realized the limits of military adventurism? It is difficult to reply very precisely on this point. Although India has no stated political or military doctrine that can be construed as a statement of intent on acquiring "great power" status, it has not hesitated to use, politically or militarily, the historical legacy of the British Empire to consolidate its territorial reach. Chinese territorial claims clearly seek historical redress from perceived wrongs by other great imperial powers and do not hesitate to use claims based on a maximalist interpretation of past Chinese territorial control (including that of past Chinese empires). As far as Central Asia is concerned, India and China would be competing on what has been an important Russian (Soviet) zone of influence. The struggle for influence would be more for natural resources and in the cultural sphere, as opposed to the clear cut strategic (military and economic) control exercised by the Soviet Union during the cold war.

For the time being, both countries profess a single-minded focus on economic growth to finance human development and ensure societal peace and national unity. Both place a high premium on peace and war avoidance. In fact, in this context, we would like to suggest a third possible element in the strategy to achieve great power status. After the stated British, French, and American doctrines of justifying imperial reach (civilizing mission, protecting interests, spreading free trade, democracy, and the rule of law), post-1945, the United States, as pointed out by the Meltzer Commission Report, 2000–2001 (appointed by the U.S. Senate to review international financial institutions), redefined the concept of national interest in a unique way: "In 1945, the United States espoused an unprecedented definition of a nation's interest. It defined its position in terms of the peace and prosperity of the rest of the world. It differentiated the concepts of interest and control. This was the spirit which created the International Financial Institutions and which has guided the Commission's work. Global economic growth, political stability and the alleviation of poverty in the developing world are in the national interest of the United States."

No matter what the distance between high principle and realpolitik, it would be difficult to envisage a modern India and a modern China emerging out of imperial domination and underdevelopment and not sharing this vision of national interest and international peace and development. They are already exercising the constituent elements of the "soft power" that has characterized America's attempt to be different from previous imperial powers. In this, they already possess a considerable advantage, like the United States did (albeit differently, as a young nation)—a civilizational depth, a culture and history, and a capacity to reinvent themselves in a manner that makes them both relevant and fascinating to the rest of the world.

However, it must not be forgotten that soft power is effective only when there is hard power to back it up. And like other great powers, India and China are undoubtedly using economic and military strength to project themselves over their zones of influence. Central Asia, the Indian Ocean, the Middle East, Africa, and South East Asia, are all terrains of competition between India, China, and the other great powers. They naturally consider some areas to be their own "backyard." India's natural backyard is South Asia, the Indian Ocean zone, the Middle East, and "near" South East Asia. For China, it would be Central Asia, its Pacific Coast, and South East Asia; it is now increasingly projecting its military and economic power in the Indian Ocean and Africa too. Behind territorial influence and the access to natural resources and markets, there is also a willingness to contest the presence of rivals.

Of course, at one level, the growing economic interdependence worldwide reduces the possibility of a classical interstate conflict of the kinds known in the past. But on another level, it is equally true that a scramble for resources—energy, water, and minerals—essential for economic growth will bring nations into conflict with each other. Wars of the future, however, may be executed through more sophisticated means such as asymmetric warfare or more non-kinetic methods such as cyber or electronic warfare. In such a scenario, sub-conventional conflict (also popularly known as low-intensity conflict or proxy war) is likely to be the norm.[8]

Central Asia—Playground for the "New Great Powers"?

One geographical region that does and will increasingly feel the impact of the rise of China and India is Central Asia. Geographical proximity, civilizational ties, rich mineral wealth, and relative absence of ideological affiliations make the five Central Asian republics (Tajikistan, Kazakhstan, Kyrgyzstan, Azerbaijan, and Uzbekistan) attractive to both Asian nations. Therefore, both perceive the region as an important ingredient for their continued economic growth (especially for various sources of energy) as well as for projecting their influence and power play. Also, since the states in the region do not have a strong ideologically bound foreign policy, it allows the republics to establish bilateral relations that serve their national interest the best. This has been aptly described as a "multi-vector" policy that "aims at counter-balancing the interests of each power, while at the same time engaging each of them politically and economically to advance its goals."[9]

The ability of China and India to cooperate and coexist in the region would impact not only them but the entire region. It, therefore, merits consideration as to how both will accommodate each other's interests and simultaneously compete to safeguard their own priorities. In fact, China and India will increasingly find themselves in what may be called "co-oplicts"—or situations where they find themselves on the same side of the fence such as in attempts to stem the rise of radical Islam or to arrest drug trafficking in Central Asia, and in issues where they will be in open competition with one another, such as over energy resources.

China and India constitute the energy heartland of the present-day world.[10] Huge energy deficits, particularly in the field of electricity, are perceived as constraining their further economic growth. Even though the per capita electricity consumption of China at 1,300 kwh is more than India's 600 kwh, both these figures pale in comparison to the per capita average in the developed economies at 15,000 kwh. It is, therefore, hardly surprising that securing reliable energy supplies preoccupies the leadership in both countries. India expects a phenomenal growth in its energy demand, estimated to be between 6–10 percent per annum during the first quarter of the twenty-first century. The power policy of India promised electricity to all by 2012, and electrification of all villages by 2009.[11] The objective for 2009 was not met and the target for 2012 also appears practically impossible given that the present total power generation of about 150 GW is woefully short of the demand that is growing by the day. For this situation to substantially change, the absolute amount of energy generated by India would have to at least double by 2020, double again over the next ten years, and be close to ten times the figure today by 2050. According to Dr Kakodkar, Atomic Energy Commission chairman, even if India's per capita energy consumption was to rise to 5,000 kwh (which would still be three times less than the current consumption figures in the United States), the country would suffer an energy deficit of 412 GW by 2050.[12] As is evident, the deficit itself would be nearly three times the current total power production.

According to another estimate provided in 2006 by the government-instituted Expert Committee on Energy, India's power needs would be about 960 GW by 2031–2032, assuming a GDP growth rate of 9 percent.[13] Since then, the global financial crisis and the consequent economic downturn have brought down the expected rate of growth of the Indian economy to 7 percent per annum. Even at this lower rate of growth, the vulnerabilities that will accompany dependence on large-scale energy import are clearly evident.

Driven by the same motivations, China too has reached out worldwide to secure adequate energy sources. China's unprecedented economic growth has made it the second largest energy importer in the world after the United States. This obviously exposes that country to vulnerabilities and in an attempt to mitigate overdependence over one supplier/region, China has scouted the entire world from Latin America to Africa to Central Asia and the Middle East in search of lucrative deals. In fact, post-Soviet Central Asia was accorded strategic importance in China's regional policy that has been described as "second only to East Asia and the Taiwan Strait."[14] Sino-Central Asian economic relations are already sealed through a multiple network of roads, railways, and pipelines for Caspian oil and gas. China has several multibillion dollar energy projects in the region, including an oil pipeline to carry crude oil from Kazakhstan to Xinjiang.

India, however, has not achieved much success in winning many deals in the Central Asian oil sector. Geographical and political barriers have prevented New Delhi, despite its strong diplomatic presence in the region,[15] from making quick inroads into the Central Asian energy market. However, India must gear up to further engage with the region given that Kazakhstan's oil reserves are believed to be on par with Kuwait's, "making it the world's major alternative energy supplier in the next ten years."[16] The country has 241 listed oil fields of which only 74 have been explored by 142 companies until now.

Given the fact that Central Asia is a landlocked region, its substantive oil and gas wealth can be accessed only through Afghanistan, China, Iran, or Turkey. Of these four, Afghanistan is ridden with instability and Iran is under international sanctions owing to suspicions over its nuclear ambitions. While Turkey caters to the Western markets, China remains one route that can be used to access oil and gas by a country such as India. In fact, since the hydrocarbons cannot safely travel to India through pipelines traversing Afghanistan and Pakistan, it might be worthwhile to examine the option of bringing the much-needed oil to northern India through the Xinjiang province of China. Already China is building pipelines to the region for its use. The possibility of extending these pipelines into India through an "energy highway" from Central Asia along Western China to northern India would make economic sense besides being of great strategic significance.

However, what are the prospects of such a development given the state of bilateral relations between India and China? Would China be interested in enabling India achieve access to resources that will propel its economic growth? What would China gain in return? These questions can be approached in many ways. First, given that Indo-China bilateral economic relations are experiencing a rapid upsurge with the bilateral trade having shot up to US$ 50 billion from only US$ 2 billion in 2000, the economic inter-linkages are growing. Of course, mistrust and suspicion over Chinese intentions remain as a result of the continuing territorial disputes and the ongoing process of military modernization. However, if China is keen on continuing to focus on its economic development then collaboration with India on building pipelines can not only be made feasible but also become another interlocking mechanism for strengthening the relationship. Given the relatively rich cash reserves that the two countries have, they can afford to jointly invest in building new export pipelines. Of course, one could well argue that China has enough financial resources to fund the pipelines by itself. But, by involving India in the exercise it would not only enhance the financial viability of the project but also have an opportunity to showcase itself as a responsible international player. The moot question, however, remains whether India and China will bring themselves together to make use of this opportunity.

Besides oil and gas, the nuclear energy sector offers great opportunities for strengthening the two countries' economic engagement with Central Asia. China is already emerging as the fastest growing nuclear power generator and if things go according to its ambitious plans, then it will be the largest producer of this energy at 130 GW by 2030.[17] Of the 55 GW of additional installed nuclear generating capacity projected for Asia, 24 GW is projected for China, 12 for India, and 12 for South Korea.[18] According to OECD (Organisation for Economic Co-operation and Development) estimates, if the nuclear activity planned over the coming decades remains on track, nuclear reactors would supply a fifth of the total electricity generated worldwide by 2050.[19]

Seen from the perspective of the ambitious civilian nuclear plans that both India and China have, their relations with Kazakhstan and Uzbekistan are particularly important given the huge uranium deposits in these nations. On January 24, 2009 India signed a memorandum of understanding with Kazakhstan, the third largest nuclear fuel producer. For India, this is significant given that lack of uranium has been hampering the functioning of Indian nuclear reactors over the last couple of years.[20] This situation arose out of two factors: first, though the country's uranium reserves are estimated at 61,000 tons and have been calculated by the DAE to be enough for 10,000 MW power generation for 40 years, uranium prospecting, mining, and milling have been relatively ignored over the last few decades. Once fast-track power plant construction started in the mid-1990s , a mismatch developed between uranium demand and supply. Second, over the decades, the uranium reserves have depleted and the ore is presently being obtained at much deeper levels than earlier. This pushes up the cost of recovery of uranium, which in the case of India is in any case high because of low concentration of uranium in the ore.[21] Therefore, helping to tide over the domestic uranium crunch is one of the relatively immediate benefits of the recent nuclear cooperation agreement to allow India to access uranium from the international market. UCIL has been keen to bid for uranium prospecting or mining in other resource-rich regions of the world.

Meanwhile, given the interest in nuclear energy for peaceful purposes in many smaller countries, India has an opportunity to export its 220 MWe reactors that would be ideally suited for smaller electricity grids. These reactors have proved their competitiveness in capital as well as unit energy costs and have a demonstrated record of safe operations. India also has the capability to emerge as a low-cost manufacturing hub for nuclear component supplies to the resurgent nuclear industry worldwide. For instance, companies such as

L&T can export nuclear reactor–building skills and/or operation and maintenance services.

A second role that Central Asia can play in the great power status aspirations of China and India is by way of becoming a sphere of influence. From the very beginning of its interaction with Central Asia in 1992, China had begun to put in place a conceptual framework of its relations, which was based on settlement of territorial boundaries, sustaining high-level political exchanges, and establishing legal treaties to underpin economic cooperation. China projected itself as a "responsible regional player and stabilizer."[22] In the creation of the SCO, China has managed to carve an important role for itself in the region. The SCO has evolved into a sort of collective security organization indulging in a range of activities such as intelligence sharing, antiterrorism coordination, and military and economic cooperation. Through the organization, China has managed to gain an influential foothold in the region even while assuaging Russia's as well as other regional countries' concerns over Chinese designs on them.

India, meanwhile, despite the historical advantage of civilizational ties with the region and the greater leverage that it was granted by Russia after the breakup of the republics, has been diffident about actively establishing its influence over the region. This is where the lack of a strategic doctrine clearly shows up. The power gap left by the collapse of the Soviet Union made Central Asia a natural arena of competition between not only countries such as China and India, but also potentially Iran as an important player in the region.

Although India does participate in some infrastructure development projects in the area and, in fact, the Central Asian petroleum and related infrastructure sectors offer employment opportunities to Indian technicians and skilled workers at a time when opportunities in the Gulf region are rapidly shrinking, India has not made full use of the potential of the region. India's skills and competence in the service sectors, ranging from IT to financial and related business services, can definitely compete with China's presence in infrastructure development. But this search for economic presence and influence requires concerted coordination between Indian business groups and the Indian State.

The scope for increasing India's economic engagement and thereby its political leverage in the region is huge. In fact, India's longstanding tradition of democracy and its relatively better record in dealing with minorities (particularly if it continues to handle domestic Muslims concerns with sensitivity) make it more attractive to the Muslim-dominated region. In fact, in this context, an area of common concern for Central Asia, India, and China would be to guard the region from becoming a fertile ground for the growth of Islamic extremism from neighboring Afghanistan. Instability in the Af-Pak region and the possibility of the militants seeking havens in sympathetic Central Asian states are equally a threat to all. As of now, Islam is a cultural issue in Central Asia, though it does have a stronger influence in Uzbekistan and Tajikistan. Religion is under state control and as a scholar has pointed out, "At the mass level, there is general interest for cultural revival which remains confined to practicing traditional customs, folk rituals and ethos rooted in Central Asian civilization."[23]

The disruptive potential of Islamic fundamentalism amongst the Uyghurs in Xinjiang province of China is a matter of great concern for Beijing. Given that the majority population of the region shares ethnic affinities with Central Asia, China does perceive a threat to its territorial unity. The Central Asian states not only provide "a ready example and inspiration" of independence but also "have been accused of providing bases and other material support for various secessionist movements in Xinjiang."[24] In fact, on this issue, there is far greater congruence in Sino-Indian than Sino-Pak interests; for instance, Pakistan is interested in exporting Saudi Wahabbism to Xinjiang in order to "encircle India with a pan-Islamic arc."[25] Xinjiang shares borders with Jammu and Kashmir. However, Islamic fundamentalism in Xinjiang should be anathema to communist China. In this respect, therefore, China and India share a common threat posed by terrorism and religious extremism as the negative fallout of instability in Central Asia will be felt by both.

Another related area of common concern would be the financing of the radical terrorism that afflicts the Af-Pak region by narcotics trade that thrives not only in Afghanistan but also in the neighboring Central Asian states. However, this was a more acute problem in the 1990s when soon after their break up from the USSR, the shattered economies of these states found an easy money-making route in the drug trade, bordering as they were the Golden Crescent (Afghanistan, Pakistan, and Iran)—the most important source of illegal drugs. Situated as the region is between Europe and Asia, it serves as a natural bridgehead for the transit of drugs, as well as for the accompanying activities of money laundering and arms trade. Drug trafficking also has implications of increased drug abuse and thereby causes health risks among the locals as also problems of law and order and socio-politico-economic instability. The link between drug trafficking and organized crime is well known. Equally well documented is the impact of these phenomena on a nation's economic and political life, leading to corruption and reduced effectiveness of governance.

Conclusion

As is evident, Central Asia offers several opportunities for India and China. The resource-rich region can provide fuel for the sustained economic growth of the energy-deficient nations, besides meeting the requirement for several other material and human resources necessary for economic development. Meanwhile the two rising nations can also offer economic assistance for infrastructure development, military cooperation for the enhancement of defense forces of the republics, and technological skills and expertise for development of strategic sectors. The fructification of these avenues would, however, depend upon not only the bilateral relations of the regional states with India and China individually, but also the bilateral Sino–Indian relationship. For both countries, the possibilities of cooperation are shadowed by the "threat" potential that each country has with respect to the other. On this front, China is widely perceived to have an advantage over India in the economic and military sphere. However, its ability to bring that advantage to bear on India is constrained by its own desire to remain focused on economic growth and also by India's leverage in terms of its international "democratic legitimacy" and its potential ability to use issues such as Tibet to mount diplomatic pressure on China.

An uneasy or conflict-riven relationship between Beijing and Delhi will create polarization among nations in the region and force them to choose one or the other. Meanwhile, a Sino-Indian relationship based on cooperation and rivalry would provide opportunities to the Central Asian states to benefit from the development potential of both nations. It would be of great benefit to India and China as well as the region and beyond if the two could help evolve a cooperative security framework in which the different regions of Asia can meaningfully develop multifaceted and overlapping relations.

Notes

1. Centre de Sciences Humaines and Centre for Air Power Studies, New Delhi.
2. P. Kennedy, *The Rise and Fall of the Great Powers: Economic Change and Military conflict from 1500 to 2000* (Waukegan, IL: Fontana Press, 1988).
3. One can see the light at the end of the tunnel, even if one is at the back of the tunnel.
4. China's White Paper on National Defence as available on China Daily website, December 28, 2004.
5. China's White Paper on National Defence as available on www.fas.org
6. M. R. Gore, "Chinese Space Programme a Threat to US?" November 5, 2009, <http://www.examiner.com/x-2547-Watchdog-Politics-Examiner~y2009m11d5-Chinese-space-program-a-threat-to-US?cid=exrss-Watchdog-Politics-Examiner> (accessed July 2010). It is a different matter that a spokesman of the Chinese foreign ministry soon thereafter denied any such intentions and reiterated China's policy to promote peaceful use of outer space.
7. For example, the Indian decision to conduct a peaceful nuclear explosion in 1974 and five more nuclear tests in 1998 despite the knowledge that it would be subjected to international opprobrium and sanctions. Or, China's response to Tiananmen Square in 1989 despite knowing that it would invite international rancor over human rights issue.
8. Indeed, India has already been a victim of such a form of warfare where China has used its "all weather friendship" with Pakistan to contain India even as Pakistan uses terrorist organizations to "bleed India through a thousand cuts."

9.   P. Stobdan, "Central Asia and India's Security," *Strategic Analysis*, vol. 28, no. 1 (2004), p. 70.

10.   In fact most of Asia constitutes the energy demand heartland. Besides China and India, Japan, Pakistan, South Korea, and South East Asia project huge energy demand.

11.   M. R. Srinivasan, "The World's Energy Resources and Needs," *Nuclear India*, vol. 39, no. 1–2 (2005).

12.   "Uranium Import Can Stave Off Looming Energy Crisis: Kakodkar," *Hindu Business Line*, July 5, 2008.

13.   Swaminathan Anklesaria Aiyar, "Nuclear Power Gives Energy Security," *Times of India*, July 20, 2008.

14.   M. K. Bhadrakumar, "India Follows China's Central Asian Steps," *Asia Times*, November 9, 2004.

15.   India was one of the only four countries permitted by Soviet authorities to maintain a consulate in Central Asia.

16.   R. Mishra, "Kazakhstan and India's Quest for Energy Security," *CAPS Issue Brief*, no. 06/09, March 15, 2009.

17.   At present, the United States is the largest producer of nuclear power at 98 GW. At present, China has 11 operational nuclear power plants that provide 1.3 percent of its total generating capacity.

18.   World Nuclear Association, <http://www.world-nuclear.org> (accessed March 2, 2010).

19.   T. Patel, "Nuclear Reactors May Supply a Fifth of Power by 2050," October 16, 2008 <http://www.bloomberg.com/apps/news?pid=20601086&sid=a9TIvUNq.ePY> (accessed February 27, 2010).

20.   Through most of 2008, the Indian power plants had to run at half their capacity levels owing to inadequate availability of nuclear fuel.

21.   Indian ore has uranium content as low as 0.6 percent as compared to some Australian, Canadian and Kazakh ores containing up to 15 percent of uranium.

22.   Bhadrakumar, n.14.

23.   P. Stobdan, "Central Asia and India's Security," p. 72.

24.   S. Singh, "China and the Emerging New Balance of Power in Asia," *Asian Strategic Review*, 1997–98 (New Delhi: IDSA, 1998), p. 227.

25.   P. Stobdan, "Central Asia and India's Security," p. 73

CHAPTER 9

Cooperation or Competition? China and India in Central Asia

ZHAO HUASHENG[1]

It would be premature to describe China and India as having close, direct bilateral contacts in Central Asia. The region does not yet figure prominently in their relationship. But it is clear that the two nations will have increasingly frequent contacts there in the future and that Central Asia will be moving higher up the agendas of both countries' policymakers. There is a common but unhelpful tendency in discussions of the Sino-Indian relationship in Central Asia to see it as competitive. China and India could indeed be competitors in Central Asia, but they could also be cooperative partners. To define their relationship in Central Asia solely as one between competitors and to look at it and plan for it from the angle of competition alone would be one-sided and unidimensional. The reverse is also true: if we define their relations as purely cooperative, seeking to understand them only from the perspective of cooperation, then that too would be neither objective nor realistic. Although China and India's common, or similar, interests in Central Asia provide a basis for possible cooperation between the two nations in the region, there also exist factors that drive them to compete with each other. The key to this issue lies in the policy choices made by the two nations, and the way ahead will be determined by whether they choose to cooperate or to compete.

Central Asia in Indian Foreign Policy

India defines Central Asia as an "extended neighbor" and must, therefore, in a sense, view itself as a neighbor of Central Asia. In principle, there is a certain rationale for this, in spite of the fact that India does not share a border with Central Asia. India is geographically close to Central Asia and it has close and longstanding historical and cultural ties with Central Asia; these bring them even closer together. Historically, traders and travelers have always moved back and forth between India and Central Asia. This has facilitated continuous cultural exchange and integration. Indian culture has had an enormous influence on Central Asia, and traces of it can still be seen today. Indian academics are largely in agreement about the importance of Central Asia for India and about India's interests there, which are to do with security, energy, and geopolitics.

Security interests figure most prominently in India's Central Asia policy. For Central Asia and India alike, the greatest security concerns—the Kashmir question, Indo-Pak relations, religious extremism, and terrorism—are closely interlinked. Hence India believes that the security situation in Central Asia directly affects its own security. This is the starting point for India's assessment of Central Asia with regard to its influence on India's security interests. Meena Singh Roy, a senior researcher at India's Institute for Defence Studies and Analyses, says,

> India has a vital interest in the security and political stability of this region. Obviously given the Kashmir angle, India cannot be walled off from the political developments which take place in the Central Asian region. Any advance by Islamic extremist groups in the CARs could invigorate similar elements active in Kashmir. For reasons dictated by geography, India's strategic concerns are tied up with the regions bordering its north and northwest. Pakistan in its northwest continues to be antagonistic towards India. Pakistan is already sponsoring cross-border terrorism in Kashmir. For India, the Kashmir issue pertains not to four million Muslims living in Kashmir Valley alone, but to the peace and security of 130 million Muslims elsewhere in India. Therefore, for India the geostrategic importance of CARs is immense. Under no circumstance can India ignore this region.[2]

With an eye first of all to its own security interests, India's principal objectives in Central Asia are (1) to prevent the growth of religious extremism and the emergence of unified politico-religious regimes; (2) to sever the links between Kashmir and Central Asian religious extremism and terrorism; and (3) to conduct strategic monitoring and control of Pakistan. India, as Rajiv Sikri, former secretary in the Indian Ministry of External Affairs, put it, "would like to encourage the development of stable and secular regimes in Central Asia, lest weakened, unstable states with centrifugal tendencies become bases for terrorist, separatist and fundamentalist elements, which could link up with counterparts in Afghanistan and Pakistan."[3]

A second, and increasingly important, interest for India in Central Asia is energy. "India...has a vital interest in the security and political stability of Central Asia, which can also be a future source of India's rising energy requirement. Accessing the oil and gas from Central Asia remains a major focus."[4] India already depends on imported energy for 40 percent of its needs, and with the rapid growth of the Indian economy, the extent of this reliance on energy from abroad is set to increase further. India's goal is to make Central Asia its new energy supplier.

As a rising power, India also harbors geopolitical ambitions in Central Asia. To put it simply, India cannot allow itself to be left out while the other powers develop a strategic presence in Central Asia. India wants to join in the "game" that the great powers are playing in Central Asia and have a space of its own in the great power set-up of Central Asia. As a Central Asia watcher put it, "India cannot afford to be left out in the cold while China, Russia, Pakistan and even the EU devour Central Asia's resources, and cement strategic bases [in the region]. India knows that they are both too late and too weak to dominate the region, but they must do whatever they can to make sure that no other state or grouping accomplishes this as well."[5] Sikri said even more clearly that "India seeks to have a firm foothold and exercise influence in Central Asia along with other great powers so that this strategically located region does not become an area dominated by forces inimical or hostile to India's interests... Aspiring to be an influential global power, India has to be a player in the unfolding 'Great Game' in Central Asia, on an equal footing with the other major players like the United States, Russia and China if it is to successfully protect its vital national interests in Central Asia."

India's Central Asia policy is to make advances through bilateral channels. In 2005, India became an observer in the Shanghai Cooperation Organisation and a member of the Conference on Interaction and Confidence-Building Measures in Asia. But these are complementary to policy, its chief concern being to further its presence in Central Asia through bilateral links.

India regards military cooperation with Central Asia as very important and has a particularly active cooperation with Tajikistan. Since 2006, there have been frequent reports that India was going to build a military base at Ayni (Farkhor) Airport in Tajikistan and these have attracted a good deal of attention. Ayni Airport is located in the suburbs of Dushanbe, the capital of Tajikistan, and was a Soviet base during the war in Afghanistan but has not been used since 1985. There is no official information available on any Indian base in Tajikistan, although a variety of reports and opinions about it have been circulating in the media and in the academic world.

Tajikistan has officially denied that such a base has been built, and there has been no official Indian acknowledgment of it, so it is not yet possible to ascertain whether India already has a military base there. However, some deductions and judgments can be made: From a strategic point of view, India would like to establish a military base in Central Asia. This would not only fit in with India's ambitions for strategic development but also have practical strategic advantages. An academic observer, Tabassum Firdous has noted that India has two objectives in seeking a small military presence in Central Asia. First, it wants to "be in close touch with the strategic situation that develops in Central Asia where the states do not have strong militaries to defend themselves or to resist disruptive forces within or from outside."[6] Second, India wants to not only scuttle Pakistan's designs for strategic depth but also force Islamabad to shift its attention to the western border.[7] If India and Tajikistan have indeed reached a cooperative agreement on the Ayni Airport question, then India will have played at least some part in the restoration of the airport. India and Tajikistan also have an agreement with Russia on the use of the base at Ayni and on wider military cooperation. India's Ministry of Defence has plans to deploy helicopters and fighters in Tajikistan.

Energy and Trade

So far as energy is concerned, India has been a relative latecomer to Central Asia, starting its exploration of the region's resources only after the mid-1990s. It is now engaged in energy cooperation with Kazakhstan, Turkmenistan, and Uzbekistan. For India, an even greater challenge than that of exploiting Central Asia's energy resources is how to transport them to India. There are few choices available to it. Building oil or gas pipelines through Afghanistan and Pakistan would clearly be extremely difficult. Even if it were possible to construct them, they could not be relied upon.

India's basic policy with respect to great power relations in Central Asia is characterized by *independence* and *balance*. "Independence" implies that India does not align itself with or move toward any one of them but strives to be on an equal footing with the other powers. "Balance" entails that India considers it to be in its interests for the great powers in Central Asia to balance one another structurally, because if any one power were to be dominant in Central Asia it would mean that India would be excluded. "Given its inherent handicaps, India cannot achieve its objectives by acting on its own in Central Asia . . . (I)n order to protect and preserve its interests in the region, India has no alternative but to closely consult and cooperate with the other major powers which have interests and a presence in Central Asia"[8]

However, India does not look on all the great powers in the same way. Deep in its strategic thinking it holds very different views of each of them. Russia has traditionally been a friend of India and has no negative significance for India in the geopolitical structure. Russia cooperates closely with India on military technology (they have common interests in non-traditional security) and encourages it to expand its dealings with Central Asia. From a strategic point of view, India, therefore, has little or no need to be wary of Russia. The United States and India have no obvious clash of strategic interests in Central Asia, the U.S. policy for Greater Central Asia coinciding with India's strategy blueprint. On the questions of Afghanistan and Pakistan, which are of particular concern for India, U.S. role is positive and crucial for India.

The significance of China for India is, however, somewhat different from that of Russia and the United States. We could say that in the Indo-Russian and Indo-American relationships in Central Asia, strategic cooperation outweighs strategic competition, whereas China and India are simultaneously cooperators and competitors, and the competitive aspect of their relationship is frequently more apparent. An Indian academic has pointed out that "Chinese expansion in Central Asia is watched very carefully in India. It is becoming clear that China is going to provide tough competition to India in both energy and trade. If Chinese expansion coincides with declining Russian influence, India will have no choice but to expand its political, economic and military capabilities in Central Asia."[9] India is actually worried that China will become the dominant power in Central Asia.

Although India is promoting its Central Asia policy strongly and considers that it has some advantages in the region—for instance, it does not have a negative historical legacy nor does it present an ideological, demographic, or territorial threat to Central Asia; it has a good deal of "soft power" and a flourishing information technology sector—its strategic presence and influence in Central Asia are both still quite weak and not nearly on a par with that of China, Russia, and the United States. It has been remarked that India has not yet fully entered into great power relations in Central Asia and that "India was never really part of any competition there."[10] The greatest weaknesses in India's Central Asia policy are the low level of trade between India and Central Asia and the fact that economic links between them are developing only slowly. India is not a core member of important multilateral organizations in the region, and this restricts the ways in which it can become involved in regional affairs as well as the depth of that involvement. It is in political relations and in the areas of culture and education that India's policy in Central Asia has been relatively successful.

The greatest obstacle for India's Central Asia policy is the lack of a direct land route into Central Asia. This is why India's strategic mission and its most pressing task is the building of an overland route to Central Asia. Unless this problem is solved, it will be hard for India's Central Asia policy to grow in scale.

China's Attitude toward India

There are many reasons for the commonly held belief that China and India are competing with each other in Central Asia. First, both states are rising powers and as such, competitiveness is part of their makeup; second, there are border and territorial disputes between them, so that geopolitically they are wary of each other, this wariness extending into Central Asia. But a more important factor that generates competition between the two is the scramble for energy and natural resources. The rapidly developing economies of both China and India require support in the shape of large quantities of energy and natural resources, thus the resources of Central Asia have become the focus of their rivalry. As a U.S. strategic analyst has said, "even as Sino–Indian relations improve, they are emerging as competitors for trade and energy markets in Central Asia."[11]

However, although competition does exist between China and India in Central Asia, it is not yet very obvious and China has no wish to compete with India. The question of Sino–Indian bilateral relations in Central Asia is rarely broached in Chinese academic circles. This is an indication that China does not regard it as important. Although China and India may well be competing in the sphere of energy resources, the quality and degree of this competition is in no respect different from India's competition with other nations, such as Russia, the European Union, and Japan, and the competition between China and India is not particularly remarkable. China's energy cooperation with Central Asia began earlier than India's, the extent of this cooperation having grown deeper than that between Central Asia and India. In addition, China has greater geographical advantages than India and has not experienced much pressure from competition with India.

China and India have broad common interests in Central Asia. Both nations have been seriously affected by terrorism and thus share common interests when it comes to antiterrorist measures. As near neighbors of Central Asia, both China and India aspire to maintain Central Asian stability. Both nations recognize the importance of the situation in Afghanistan for security in the region, hoping they can help to find a solution to the Afghan problem. Some commentators are of the view that China is a negative factor in India's becoming involved in solving the Afghan issue,[12] but that is not an objective view and China would not and could not reject India's efforts in Afghanistan. In fact, China's relations with India over the Afghan question are also cooperative in nature.[13]

China takes a constructive view of its relations with India in Central Asia. In fact, China already sees India as a cooperating partner in the region. Together with Mongolia, Pakistan, and Iran, India is an observer to the Shanghai Cooperation Organisation; although observers are neutral, in practice they are seen as cooperators. The regulations for SCO observers stipulate that recognition of the organization's goals and its principles and activities is a prerequisite. In addition, initial trilateral strategic conversations among China, Russia, and India have already taken place, and the presidents and foreign ministers of the three states have held meetings touching on cooperation in Central Asia. With the establishment of the above framework, Sino-Indian relations in Central Asia fall into the structural category of cooperative partnership. As an observer, India is part of the SCO cooperative framework and is permitted to participate in cooperation in all of the SCO's areas of activity, including energy cooperation. This also covers Sino-Indian cooperation.

However, at the present stage, Sino-Indian cooperation in Central Asia occurs largely within a multilateral framework, mostly involving problems at the macro level. There is no bilateral cooperation and they are not cooperating on any specific projects. If Sino-Indian cooperation in Central Asia is to become more specific and gain greater depth, and if substantial bilateral cooperation is to be developed, a range of factors will need to be taken into account or rather will determine what occurs. These factors include the requirements and degree of urgency of specific interests, the effect they have on the region, and other related aspects. Cooperation in Central Asia between China and India could be of great significance for relations between the two nations, with potential strategic value, but deeper cooperation between them in the region will require greater strategic maturity, that is, mutual strategic understanding and a deepening of trust. Sino-Indian links are currently developing in that direction, but enough progress has not yet been made. With regard to specific interest requirements, at present the two nations objectively see no urgent need for bilateral cooperation. Where there are no obvious interest requirements, even if there is some bilateral cooperation, it may not be active for lack of specific content and sustained development would be difficult. In any cooperation between China and India in Central Asia, the Pakistan factor has to be taken into account. Traditionally, Pakistan has been China's strategic partner, and Sino-Indian cooperation must not make Pakistan feel insecure, or else it would damage trust between nations in the region.

India's military presence in the region is an important issue. Even though neither India nor Tajikistan has officially acknowledged the existence of an Indian military base in Central Asia, it is probably true that India wants to build up its presence there. The development of an Indian military presence in Central Asia is to a great extent aimed at Pakistan and has a clear geopolitical purpose, one that is detrimental to the stability and balance of South Asia. From an antiterrorism perspective, the role that a military base could play is limited. India's military presence in Central Asia will further intensify the militarization of that region and provoke the great powers into military competition. This is not the right direction for the development of Central Asia.

Conclusion

China and India should and need to cooperate in Central Asia. Cooperation is the proper political direction for them to take and will benefit both nations as well as others in the region. It is not unusual for competition to exist in certain fields, but this should be treated normally and constructively, according to the current rules of the game. It is vital to banish excessive competitiveness from Sino-Indian relations, for any such element would result in a distortion of the relationship. China and India have no need to see each other as competitors simply because they are both on the rise. Each should strive to develop itself and should not design its objectives with reference to the other, certainly not in order to crush the other. Objectively, the idea that China and India are in competition should not be too strong, for China does not have a strong sense that it is competing with India, whether at a national level or regional level, including in Central Asia.

Notes

1. Director of the Center for Russian and Central Asian Studies, Fudan University, Shanghai, Popular Republic of China.
2. M. Singh Roy, "India's Interest in Central Asia," *Strategic Analysis*, vol. 24, no. 12 (2001).
3. R. Sikri, "Behind Oil and Gas: India's Interests in Central Asia," June 29, 2007, <http://opinionasia.com/IndiasInterestsinCentralAsia> (accessed February 26, 2010).
4. A. Patnaik, "Central Asia's Security: The Asian Dimension" in R. R. Sharma (ed.), *India and Emerging Asia* (New Delhi: Sage, 2005), p. 221.
5. P. Frost, "Great Decision Analysis: India in Central Asia," FPA Features, May 29, 2008. <http://www.fpa.org/topics_info2414/topics_info_show.htm?doc_id=687617> (accessed February 27, 2010).
6. T. Firdous, "India and Central Asia: Vanishing Distances," <http://www.centralasia-southcaucasus.com/index.php?option=com_content&task=view&id=51&Itemid=46> (accessed February 27, 2010).
7. Ibid.
8. R. Sikri, "Behind Oil and Gas: India's Interests in Central Asia."
9. G. Sachdeva, "India's Attitude towards China's Growing Influence in Central Asia," *The China and Eurasia Forum Quarterly*, vol. 4, no. 3 (2006), p. 34.
10. Ibid., p. 24.
11. E. Wishnick, *Russia, China and the United States in Central Asia: Prospects for Great Power Competition in the Shadow of the Georgian Crisis* (Strategic Studies Institute, February 2009), p. 27.
12. M. Hanif, *Indian Involvement in Afghanistan: Stepping Stone or Stumbling Block to Regional Hegemony?* (Hamburg, GIGA working papers, April 2009), p. 22.
13. On October 27, 2009, in Bangalore, India, another meeting of the foreign ministers of China, Russia and India was held, at which such issues as the situation in Afghanistan and regional security were discussed.

PART III

Chinese and Indian Economic Implementations from the Caspian Basin to Afghanistan

CHAPTER 10

Scramble for Caspian Energy: Can Big Power Competition Sidestep China and India?

P. L. DASH[1]

The post-Soviet developments in and around the Caspian Sea have been so incredibly swift that observers and analysts find it difficult to keep abreast of all nuances of changes year after year. When the entire Caspian theater shifted possession from two owners to five owners of the sea, the arithmetic of everything surrounding the Caspian suddenly changed. Some of these developments such as ownership dispute over the Caspian, the legal status of the sea, possession and access to seafaring and exploitation of resources, building of the Navy for each independent state, and other similar issues are quite baffling simply because nearly two decades of negotiations have yielded few tangible results to resolve mutual bickering. Besides these burning issues the Caspian Sea found itself in the vertex of an unprecedented geopolitical competition surrounding its hydrocarbon reserves. These competitions have become so intense over the years that after the successful construction of the Baku-Tbilisi-Ceyhan (BTC) oil pipeline and the parallel Baku-Tbilisi-Erzurum (BTE) gas pipeline, a new gas connectivity called Nabucco between five countries—Turkey, Bulgaria, Romania, Hungary, and Austria—has taken shape. Its doors are open for others to join. This has considerably sharpened the ongoing regional geopolitical competition on a scale never seen before.

Both these western-sponsored pipelines have, possibly intentionally, ignored two potential and emerging trends. First, the galloping growth of China and India, their growing economic prowess, politico-diplomatic posturing on the periphery of Central Asia/the Caspian Sea, and the geostrategic equations associated with recent developments in the Af-Pak region have not been adequately addressed. Had all these been taken into account, a priority gas pipeline would have merited urgent construction from the Caspian Sea/ Central Asian region to China and India. However, this has not happened. Among the numerous pipeline projects, the materialized ones are BTC and BTE, which are already operational, and Nabucco, which is firmly proposed. Eastward looking pipelines are entirely absent, although economic growth potential of the east does not match adequately with its energy reserves because energy consumptions in China and India heavily outweigh indigenous production.

Second, the actors in this region have been ignoring and bypassing the Russian Federation. Russia remains an energy giant in Eurasia—a fact that no one can ignore, but the BTC-Nabucco projects have deliberately sidestepped Russia.[2] In the case of Nabucco, the planners have not even clearly indicated where the gas would be supplied from. Ignoring Russia and her huge maneuvering potential in her Eurasian underbelly is a fatal mistake Nabucco planners have made. Third, Chinese activities in the realm of energy politics have been ignored, primarily in Kazakhstan and secondarily in Turkmenistan, such a scenario may not allow much of the Caspian gas from the eastern flanks to flow westward. The Chinese sensibilities and India's pragmatic requirements are two factors that regional oil and pipeline politics ought not to ignore. Finally, the division that Nabucco is likely to create among the five Caspian states will accentuate the scramble for energy and invigorate geopolitical jockeying among natives and nonnatives alike, thereby encompassing many powers in a great game hitherto not witnessed. This chapter briefly addresses the competition for Caspian energy by analyzing the two pipelines BTC and Nabucco, while simultaneously focusing on the energy requirements of China and India and how possibly they could be met from the Caspian oil and gas supply.

Baku-Tbilisi-Ceyhan

When the idea of laying pipelines from Baku in Azerbaijan via Tbilisi in Georgia to Ceyhan in Turkey was first mooted, it was mocked at by experts on many counts and criticized by pessimists as impractical. The first argument was that the Caspian basin did not have enough energy reserves to deserve an exclusive pipeline. In Soviet years Azerbaijan was supplying Caspian oil through two Soviet era pipelines running Baku to Supsa via Georgia, and Baku to Novorossiisk via Russia. The BTC was proposed to change Azerbaijan's energy profile forever by opening a new market outlet to Turkey. It was for the first time that Westerners would get access to the Caspian energy resources—an access that even Hitler had failed to get during World War II. For the first time the BTC was planned to completely bypass Russia, which seemingly lost Azerbaijan to the West.[3] For the first time too it was destined to ignore another Caspian power—Iran, the arch enemy of Washington in the Caspian basin. For the first time the Caspian was connected with the Mediterranean by an oil pipeline, thanks to Western political commitment. For the first time economic consideration was overplayed by geopolitical factors and for the first time world leaders had evinced direct interest in the creation of the BTC. Even U.S. president Bill Clinton personally endorsed the pipeline and promoted the BTC on many occasions.[4]

Why were Baku, Tbilisi, and Ceyhan the puzzles that nonplussed analysts and observers for a while despite all these "firsts." When the pipeline was so designed as to pass close to many ethnic hotspots in the Caucasus, such as Nagorno-Karabakh, South Ossetia, Chechnya, and the Kurdish area in northern Turkey, but sidestep them all cleverly, it naturally raised many questions that remained unanswered. However, as years passed by, the U.S. geopolitical considerations became clear. The project was designed to firmly ally with Azerbaijan, befriend Georgia, and advance the interests of Turkey as NATO's strongest representative in Central Eurasia. This was why the Western choice fell on the BTC. In 2002 when the construction of the BTC started, everybody was sure about the West's commitment to this project. And three years later, in 2005 much against the pessimists' doomsday predictions for the BTC, the pipeline pumped oil to Turkey. Parallel to the oil pipeline, a gas pipeline was also laid to connect Baku in Azerbaijan with Erzurum in Turkey, from where future Nabucco would take off.

The initial project cost of the 1,730-kilometer-long (2,000 kilometers by other estimates) BTC was estimated to be US$ 3 billion dollars.[5] Subsequently, the project cost was raised to US$ 3.6 billion and it was planned to carry a million barrels of oil a day, which is a little over 1 percent of the world's daily oil consumption.[6] The determination for geopolitical gains was so sterling that economic viability and cost escalation did not matter at all. Further estimates suggested that by the time the BTC was commissioned, it cost a little more than US$5 billion.[7] Of the 2,000 kilometer stretch, 550 kilometers passed through Azerbaijan territory, 250 via Georgian land, and the rest 1,200 on Turkish territory, thereby implying that Turkey will be the largest beneficiary of the BTC and the BTE.[8] The cost-effectiveness of the project thus mattered little in terms of economy. It was a political victory, a success story par excellence, the offshoots of which may navigate farther than Erzurum or Ceyhan. The BTC and the BTE forever changed the contours of oil politics in the Caspian basin. Russia and Iran, which traditionally transported Caspian oil to various market destinations, lost their monopoly the moment the United States ensured that the BTC would prevail and function uninterrupted. It was the scramble for energy by these three powers—Russia, Iran, and the United States—that determined the fate of Caspian geopolitics the way it stands today.

Genesis and Growth of Nabucco

Much before the BTC was commissioned, way back in February 2002, Austrian firm OMV and Turkish firm BOTAS began preparing a gas pipeline project that turned out to be the genesis of Nabucco. Three months later, in June 2002, three other oil companies—MOL of Hungary, Bulgargas of Bulgaria, and Transgas of Romania—expressed their desire to join the duo, and in October 2002 all of them signed a cooperation agreement in Vienna that laid the foundation of Nabucco. After the signing ceremony of the cooperation agreement, all five partners visited the Vienna state opera to see Giuseppe Verdi's play "Nabucco," based on the Babylon and Jerusalem of sixth century BC, first staged in March 1842. All five members unanimously named their dream project after that play, thereby giving the project an unusual name.

A year later, some speculative and some pragmatic initial studies were undertaken to assess Nabucco. In December 2003, the European Commission was ready with an unspecified award of 50 percent of the total cost of feasibility study of Nabucco that included technical details, market analysis, economic and financial implications, and other aspects. After a gap of one and a half years, on June 28, 2005, the five partners signed a joint venture agreement that they had to formalize a year later on June 26, 2006 with a ministerial statement on Nabucco gas pipeline. Subsequently in February 2008 Germany's RWE became a shareholder of the Nabucco consortium, thereby making Nabucco a six-party gas venture to take off into the unknown future because it was planned to be laid on a very uncertain and treacherous terrain.

Azerbaijan was the first gas-rich country with which Nabucco inked its first ever gas procurement deal on June 11, 2008 to supply gas to Bulgaria. Ilham Aliev, the president of Azerbaijan, had then stated that his country would double gas supply to Nabucco in five years. As the Nabucco ball rolled on a definitive path of charting out a pipeline project, the European Bank for Reconstruction and Development as well as the European Investment Bank expressed their willingness at a Nabucco summit in Budapest on January 27, 2009 to finance the project. The very next day, the European Commission proposed a 250 million Euro package for funding the Nabucco project. Thereafter Nabucco was on the agenda of all top ministerial-level meetings held between January and June 2009 until finally the five partners inked its founding, international document on July 13, 2009 at Ankara in the presence of U.S. senator Richard Lugas and U.S. special envoy on Eurasian energy Richard Morningstar. Representatives of all other parties, including the European Commission that had agreed to patronize the project and extend financial support, were present too.

The 3,300-kilometer-long gas pipeline from Erzurum in Turkey via Bulgaria, Romania, and Hungary would connect with Austrian natural gas hub at Baumgasten an der March. Of these 3,300 kilometers, 2,000 will be laid on Turkish territory followed by 400 in Bulgaria, 460 in Rumania, 390 in Hungary, and 46 in Austria; of these, four will cover station facilities. Far too easy it may seem to build these connects, but far too difficult it would be to procure gas to fill the pipelines. However, Nabucco will eventually provide a cross-country connect and a cross-country market. The trans-Caspian gas pipeline and the South Caucasus gas pipeline will be joined with Tabriz-Erzurum pipeline, ultimately connecting major consumers with major producers in the Balkans, East Europe, and Caspian/Caucasian sectors.

Construction of Nabucco will commence in 2010 to end in 2014 at a proposed cost of 7.9 billion Euros. Nabucco proposes to procure gas from major gas producers such as Azerbaijan, Turkmenistan, Kazakhstan, Iraq, Iran, and Egypt. Poland, although far away from Nabucco, has expressed willingness to link with it, apparently to minimize her energy dependence on Russia. Big producers such as Qatar may subsequently join. As Nabucco transformed from a mere opera name to an important international gas pipeline, ambitious in nature and rich with promises of finances flowing its way, final configurations of where the gas would come from and where the pipelines would be laid are becoming all the more urgent issues of decision making for the planners. Clearly, Nabucco is going to be the most important, longish, and third important pipeline, in addition to the BTC and its parallels the BTE, built in the post-Soviet decades. Its completion would further hurt the injured ego of Russia in the world gas market. As time pass by, the countries of Russia's southern underbelly would find it increasingly harder to deal with Russia, not necessarily because it has tried and succeeded to regain its foothold in Caucasian affairs after the 2008 August war in Georgia, but surely because Russia has a tangible interest in the region as a gas giant that can hardly be ignored.

Turkey has become one of Russia's largest trade partners in recent years. A change of government in Bulgaria has taken away the sheen from Russia's Slav sentiments. Gas supply to Ukraine and via Ukraine to several other European countries remains a knotty issue to be addressed. The Caspian countries have not been able to bury their hatchets because Russia has been pursuing an adroit diplomacy in the region by aligning with oil and gas–rich countries to negate the U.S. influence in the region. A bit too rapidly too many things are happening to make the intense geopolitics of the region around oil and gas a very sticky and sensitive issue to deal with. And a way has to be found to deal with Russian sensitivities and address the Russian national interests that have a far greater bearing on the development in the region than the U.S. presence. None of these countries in Russia's southern underbelly, including

Turkey, can match Russia's economic strength, political weight, diplomatic ability, and undiluted tenacity in pursuing a course in defense of its national interest.

The pragmatic picture thus far of the region relative to oil and gas supply is bizarrely lopsided. All pipelines in the south from the oil-rich Persian Gulf and West Asian regions look westward to Europe. Similarly all Russian pipelines in the north from central Russia to Siberia, including those from the Caspian littoral states, are oriented to westward export. It is for the first time that the western oil companies have gained access to the trans-Caspian oil and gas resources without Russia being included. Pipelines, looking eastward and supplying energy to the thirsty economies of the south and Southeast Asian countries are absolutely absent. Extrapolating pipelines from energy-rich Kazakhstan and Turkmenistan are also absent in the eastward direction. They are only being proposed, or hardly; a couple of them from Kazakhstan to China have materialized.

As the forward march of China and India began at the threshold of the millennium, speculations arose about their real energy requirements and how the energy-rich regions and countries of the world could address the growing needs of these two demographic giants. Thus the question of how to diversify the transportation network came up. However, whatever has materialized so far are again in the westward direction, looking at Europe's stagnating economy (growing at a rate of 1–2 percent) and ignoring the leaping economies of Asia (growing at an average annual rate of 7–8 percent). This issue cannot be kept pending for far too long. And traditional ways of transporting energy requirements of India and China through conventional means—tankers, railways, and ships—ought to pave way for pipeline connectivity from the sources of production to the points of consumption.

Factoring China

As the BTC materialized and Nabucco formed, the Caspian/Central Asian region witnessed many actors—all vying for a strategic foothold with a view to gaining access to the hydrocarbon resources. The arithmetic of the basin swiftly changed. The first post-Soviet arithmetic that the Caspian Sea encountered was an increase in the number of owners from the pre-Soviet two to a post-Soviet five. All further postulates with regard to determining the legal status of sea, division of the sea shelf, fixing the access zone for each country, dividing the hydrocarbon resources, building native armies, air force, and navies, and all such factors encountered different arithmetics. The Caspian basin in particular became an arena of intense geopolitics at the advent of the millennium. And new emergent powers in the east—such as China, India, the

Koreas, and other East Asian states—yearned to have their share of the energy pie from the Caspian/Central Asian region.

China was the first to visualize the importance of hydrocarbon reserves in the Caspian that it badly needed for her own economic growth. If the United States could patronized the BTC and the BTE, why not China in neighboring Kazakhstan? China could easily target Kazakhstan—a Caspian power with an endowment of enormous hydrocarbon reserves—and propose eastward-looking pipelines from western Kazakhstan to mainland China. Twenty years ago, when the Soviet Union collapsed, no one believed that China would emerge as an economic giant, but today China is a reality that no one can ignore. China is the world's third largest economy.[9] From a non-importer of energy in the 1970s, China today is the second largest importer of oil and gas. In the era of globalization, China's regional isolation is impossible; today Tiananmen and Tibet are as much a nonissue as democracy and rights violation inside China. Memories of the Cultural Revolution are buried in the past. What is visible in China at 60 is her success story, not her frailties or failures.

As China celebrated her sixtieth birthday on October 1, 2009, it demonstrated a military might it had never shown earlier: 52 types of new weapon systems (90 percent of which were paraded for the first time), five types of missiles (including ballistic, intercontinental, and nuclear), and 151 warplanes (ranging from its most advanced J-10 and J-11 fighter jets to AWACs, bombers, and aerial tankers).[10] In a two-hour pageant 200,000 servicemen and women from 56 regiments symbolizing the country's ethnic numbers paraded to demonstrate China's military might to the world. Equally visible is Han Chinese nationalism all round, but much less of innovations and creativity in science and technology. China's economic growth has given her a sense of pride, and a bit of arrogance. It was clearly demonstrated in the July–August disturbances in Xinjiang and in the 2008 crackdown in Tibet. What matters, however, is whether China can be ignored from the ongoing scramble for oil in the Caspian. Its adroit handling of its policy with Kazakhstan and Russia demonstrates that China could no longer be ignored.

Some comparisons of emergent China are not entirely misplaced here. In 1978, China accounted for just 1 percent of the world economy, today it is 5.5 percent. Although China's economy remains one-fifth the size of the U.S. economy, the United States has failed to keep a balanced trade position with China since 1983 and this trade imbalance has been growing exponentially and consistently over the years from US$ 6 billion in 1985 to US$ 203 billion in 2006. The rise in U.S. trade deficit with China has obvious outcomes: over two decades displaced productions have cost the United States 2.2 million jobs that it could have otherwise maintained.[11] Certainly China is a source of worry for the Western architects of Caspian pipelines, be it the BTC or

Nabucco. With growing Chinese entry into Kazakh and Turkmen oil fields, the hiatus will continue and hopes will fade to get either oil to the BTC or gas to the BTE or Nabucco. If China could provide a reliable next-door market for Kazakhstan and Turkmenistan, demand for these two countries to supply oil and gas to farther places will seem remote in the long run. This would inevitably result in construction of eastward pipelines in the direction of China.

Kazakhstan seems to be China's energy partner of the future. China has already been deeply involved in many energy projects in Kazakhstan, investing US$3 billion in energy projects that would cater to its future needs. All pipelines, transit hubs, hydroelectric power stations, and oil storehouses are heading in the direction of Xinjiang Autonomous Republic. As a result, what was lying supine in China's northwestern remoteness has gained prominence in development with the rapid growth of road and rail infrastructure. The Urumqi-Kashgar rail line, the Urumqi-Druzhba rail line, and the road across Tibet are testimony to the shape of future development in that region. As road, rail, and communication infrastructures take their place, interstate and intrastate mobility will correspondingly increase. China has been looking to an alternative rail route to Central Asia to connect Urumqi with Andijan as part of the Eurasian continental bridge that would connect through rail the eastern flanks of China to the Atlantic coast by rail via a Baghdad-Berlin line. This will certainly facilitate movement of oil and gas from the Caspian/Central Asian region to even unimaginable destinations in the foreseeable future.

China is more proactive in its contiguous region than the United States had been in Turkey or through its own military presence in Central Asia. While the Eurasian continental bridge holds sway, China has been looking northward to Russia and Kazakhstan, westward to Uzbekistan and Turkmenistan for sustained energy supply and possible engagement in energy infrastructure projects, and southward to Pakistan for a quick and reliable strategic access to warm waters as well as for oil supply from the Persian Gulf region. It has been nurturing Pakistan for years by investing time, money, technology, and expertise to semicircle India from the west. It has built a deep-sea port at Gwadar and connecting roads from Kashgar via Pakistan-occupied Kashmir to that port. Its nuclear designs with Pakistan are all the more evident in the proliferation activities of the disgraced nuclear scientist A. Q. Khan. All these efforts are to ensure sustained energy supply—oil, gas, and nuclear—from wherever it is available.

With regard to the scramble for Caspian oil and gas, the Chinese focus is centered around three Caspian countries—Russia, Kazakhstan, and Turkmenistan. The focus is diverted to a negligible extent to Uzbekistan too. Even these countries in the periphery of China cannot afford to ignore the huge market it provides for energy. In this context, the Sino-Russian oil deal of February 2009 deserves mention. This was a major development in the Sino-Russian energy cooperation. As per this deal Russia would supply oil to China in exchange of a loan that would facilitate exploration and marketing of Russian oil. A US$ 25 billion Chinese credit will be distributed 15:10 to Rosneft and Transneft, Russia's state-owned oil firm and pipeline firm respectively, in exchange of 300,000 barrels of oil a day for the next twenty years. In an effort to diversify the source of its huge oil import, China, the world's second largest importer of oil, has struck this deal to minimize her dependence on Gulf oil. In the eastern flanks, Russia has been looking for markets in Japan and China for its Siberian export. Similarly, China has been looking at Russia, Kazakhstan, Turkmenistan, Africa, and some South American countries as alternative sources of oil import to ensure its energy security.[12]

Australian prime minister Kevin Rudd was right when he said, "There is no simple one-line answer to the question of how we should seek to engage China. It's a huge country with complex global, domestic and historical currents that influence its current policy decisions. But one key is to encourage China's active participation in efforts to maintain, develop, and become integrally engaged in global and regional institutions, structures and norms."[13] China has thus engaged not only Russia but also Kazakhstan in a substantial way. The CNPC has four major projects in China running into billions of dollars: (1) The project for transporting oil by rail from Kazakhstan to Xinjiang, worth US$ 300 million; (2) Construction of the West Kazakhstan-China pipeline project, worth US$ 4.3 billion; (3) Widening the diameter of gas pipeline between Almaty and Urumqi, worth around US$ 2 billion; and (4) Ili River hydroelectricity project, worth US$ 250 million.[14]

By invoking civilizational links with Turkmenistan from the Silk Road days, China has entered the Turkmen gas fields. Two exclusive licenses have been issued to China for exploration and production of gas in Turkmenistan, when the law was stringent for others. As a result, a gas pipeline to China is being constructed. Turkmenistan's reserves of 7 trillion bcm of gas and 4.4 billion barrels of oil are enough for regional energy security, and the country adopted a law in August 2008 to facilitate joint exploitation and marketing of which China is likely to take advantage. Subsequent to this law the EU, Russia, Iran, India, China, and Pakistan have joined the scramble for Turkmen gas.

Emergent India

Although China's incredible economic growth since the early 1980s has dwarfed India's economic success, the India of the twenty-first century is another success story of Asian economic reforms. India started her economic reforms a decade later than China; ever since market reforms were introduced in 1991–1992 to part with the license raj of yesteryears, India has quickly moved forward in two directions: the creation of a knowledge economy, particularly in the emerging IT sector, and the liberalization of her economy for foreign investment. As a result, India has gained economic success year after year and has sustained an economic growth of 7–8 percent over a decade minus the years of recent economic slowdown. In the sustaining of this growth India's energy demand has soared rapidly. The other impediment that came in the way was the breakdown of the Soviet channel of oil supply, first, in 1991 as a result of the collapse of the Soviet Union, and, second, in 2003 with the fall of Saddam Hussein in Iraq from where India was getting Soviet oil supply for years on the strength of an Iraq-Soviet "oil for arms" swap deal.

Despite India seemingly having a significant presence in world affairs, its quest for energy security continues to flounder. Its efforts at getting gas from Kurmangazy in Kazakhstan has fallen flat with China winning the bid. The North-South transportation corridor has not picked up yet, pipelines from Central Asia and the Caspian region remain a remote daydream. All these failures do not, however, portray India negatively. In the Caspian and Central Asian sectors, India's efforts to ensure energy security are noteworthy. In the past two decades it has targeted four Caspian powers to gain access to Caspian hydrocarbon reserves and consequently to get oil and gas. However, unlike those of China, all these efforts are more reactive than proactive. As a result, years after the initiatives were taken, no concrete outcome has come India's way.

Of all the Central Asian/Caspian countries, India looked positively at getting Kazakh and Russian oil and Turkmen and Iranian gas. Primarily it aimed at partnering Turkmenistan in its quest for energy and favored two major pipeline deals: Turkmenistan, Afghanistan, Pakistan, and India (TAPI) as well as Iran, Pakistan, and India (IPI). Due to inherent political instability prevailing in the Af-Pak sector, and Iran not being in the good books of the United States, both projects stand unimplemented and their fate hangs in the balance. Turkmenistan is ready to supply gas to India from its Dauletabat field whichever possible way it could, but these ways are not getting cleared. Although the first gas consortium was set up in 1996 to work toward a trans-Afghan pipeline in the direction of India via Pakistan, military and political instability has impeded or virtually stalled the project. In a review effort in December 2002, Turkmenistan, Afghanistan, and Pakistan stressed the importance of a pipeline in this area and signed another agreement that India subsequently joined in 2008 to buy Turkmen gas. This deal has ever since been marred by continuing troubles in Afghanistan and political instability in Pakistan. The IPI has been weltering with the unfolding developments in the gas price formula: Iran asking for more and India not agreeing to pay more. The United States also has been trying to stymie IPI and favor TAPI. "Despite the de facto suspension of India's participation, Islamabad and Tehran have managed to agree on the gas price formula at the end of 2008 and are making optimistic noises about pushing ahead with the project."[15]

Iran's possible participation in the Nabucco project to supply gas to that pipeline will distract Iran from the IPI, making it vulnerable to the vicissitudes of emerging trends. Growing tensions in Baluchistan—Pakistan's gas-producing province—will divert Pakistan's attention from both TAPI and IPI, thereby making difficult any tangible prediction about their final implementation. This is a bizarre situation for India to handle. In any case to write off India from the Caspian and Central Asian energy scenario is premature. India would continue its efforts to gain a foothold on the Caspian energy platform by fostering good ties with all Caspian states.

India's growing energy needs are currently not met from the Caspian supply. India has adopted a multipronged strategy of energy procurement by not putting all its eggs in the Russian basket as it did in the past. It has diversified the geography of involvement by diverting its resources to participation in the Sakhalin projects to get Russian oil supply from the Far East. It has been pursuing various other options in Turkmenistan, Iran, and Kazakhstan to get gas supply from the former two countries and uranium from the later. It has also kept its options open with regard to TAPI and IPI, while exploring the possibility of extending those pipelines to India's northeast and from there to Myanmar. The reported discovery of about 36.5 trillion cubic feet of gas and 4.6 billion barrels of oil in northern Afghanistan by the U.S. Geological Survey and Afghan Ministry of Mines in March 2006 has raised hopes of Afghanistan actively joining TAPI and increasing its prospect of serving as a transit country for transportation of gas to Pakistan and India.[16]

Thus India is very much afield in the pipeline politics of Eurasia, particularly in its neighborhood. As the demand for energy grows manifold, India's engagement in the Caspian and Central Asian region is likely to be intense in the years to come. India was the second fastest growing economy of the world in 2003 and in July [...] UN ranked India as the world's tenth largest economy and the four[...] he third after Japan and China. Compared t[...] rofile is starkly poor. The Tata Institute of [...] ch a profile of continuing progress will dr[...] il up to 80 percent and on gas up to 77 pe[...] confirmed. To think of such a country ge[...] ics of oil is a misplaced notion.

Many actors and several factors will play their role in the emerging geopolitical contours of energy politics in Eurasia in the foreseeable future. The inevitability of ensuing energy security for the rising economies of China and India is a fact the Caspian and Central Asian countries cannot ignore any more. Political equations may change dramatically, dropping Iran from the bad books of U.S. diplomacy and paving way for IPI or even TAPI. In the scheme of things to come Iran holds the key to the future geopolitics of oil, primarily because of its central location: close to the Central Asian countries, this Caspian power is capable of providing access to both the Caspian Sea and the Persian Gulf and having road and rail network access in all directions. While TAPI and IPI may be on the regional geopolitical hold, what matter are the multiple transportation options for India—the sea-land-rail link through Iran's Chabahar and Bandar Abbas ports. This route as part of the north-south corridor has not taken off, yet it provides by far the safest option for India to access the Caspian and Central Asian energy resources because Iran is the only country that has north-south access to the sea and east-west access to land. While big powers have succeeded in laying the BTC and planning Nabucco, the small and emerging powers are in quest for tangible route options that would address their needs and aspirations. Until that is achieved, Indian and Chinese tryst with geopolitics of oil will continue to overwhelm their efforts in international posturing.

Indo-U.S. endeavors to better the Afghan situation with the help of Pakistan will pit them in one block to favor TAPI against Russia, Iran, and India who may favor, amidst turbulence in Pakistan, the IPI and China, cutting across the region in the middle of two groups to obtain a road access from Kashgar to Gwadar, thereby ensuring her supply of energy from the Persian Gulf to its undeveloped western flanks. China will thus have three supply lines: one from Gwadar to Xinjiang via Pakistan, many others from Kazakhstan to Urumqi, and the third from Russian western Siberian fields to Daquing. In comparison, India has two options that it uses today: get Caspian and Central Asian as well as Russian oil and gas supply via Iran and exploit further the possibilities of getting energy supply from Russia's Sakhalin by sea. Pipelines are a distant pipe dream for the time being.

Notes

1. Professor of Eurasian Studies, University of Mumbai.
2. A. Samozhnev, S. Zykov, M. Tchkanikov, Kyda Jobyet "Nabocco," *Rossiskaya Gazeta*, July 14, 2009, p. 5.
3. Mikhail Rostovskii, "Iskushenie Neftyiu: Kak Rossiia poteriaet Azerbaijan," *Rossiia i Musul'manskii Mir*, no. 11 (2005), pp. 95–97.
4. B. Shaffer, "From Pipedream to Pipeline: A Caspian Success Story," *Current History*, October, 2005, p. 343.
5. M. Malek, "Politika Bezopastnosti na iuzhnom Kavkaze: Ocnovnoi Krug Problem," *Tsentral'naia Aziia i Kavkaz*, vol. 30, no. 6, 2003, p. 16.

6. B. Shaffer, B. Shaffer, "From Pipedream to Pipeline: A Caspian Success Story," p. 343.

7. Padma L. Dash, *Caspian Pipeline Politics, Energy Reserves and Regional Implications* (New Delhi: An Observer Research Foundation Publication, Pentagon Press, 2008), p. 42.

8. K. Celik, K. Cemaletin, "Azerbaijani Petroleum: Past and Present," *Eurasian Studies*, no. 16 (1999), p. 120.

9. M. Liu, "Life Begins at Sixty," *The Times of India*, September 28, 2009, p. 16.

10. *The Times of India*, October 2, 2009, p. 1.

11. I. Shariff, B. Mak Arvin, "China in Global Economy: A Threat to the United States?" *World Affairs*, Vol. 11, no. 4 (2007), pp. 99–100.

12. *Economic Times*, Mumbai, February 19, 2009, p. 9.

13. Cited in M. K. Bhadrakumar, "Engaging China as a Friendly Neighbour," *The Hindu*, April 10, 2008.

14. L. Muzaparova, "Sino-Kazakh Energy Cooperation: Assessment of Potential and Direction of Development," in *Kazakhstan and China: Strategic Partnership for Development*, International Conference Proceedings, March 2, 2006, pp. 21–22.

15. A. Zulkharneev, "Iran's Energy Interests in the Caspian: The Window of Lost Opportunities," Security Index, vol. 15, no. 3–4 (88–89), 2009, p. 61.

16. S. Blank, "Afghanistan's Energy Future and Its Potential Implications," <http://www.eurasianet.org/departments/insight/articles/eav090306.shtml> (accessed March 1, 2010).

CHAPTER 11

Comparing the Economic Involvement of China and India in Post-Soviet Central Asia

Sébastien Peyrouse[1]

The assertion that Central Asia has become a field of economic competition between China and India is based on a prospective approach. The potential of the two countries is indeed to be competitive, but economic realities show that it is not the case at the end of the first decade of the twenty-first century. Comparisons between India and China in Central Asia, though legitimate in the geopolitical logic of power projection, is less relevant in the economic realm, the comparison being valid only over the medium or long terms, 2020 to 2030. Indeed, for the time being, except for hydrocarbons, where the Indian and Chinese companies have already come to terms, China largely dominates all the other areas. A comparison of trade flows shows that the total trade between China and Central Asia exceeded 18 billion euros in 2008, whereas between India and Central Asia, it was only 247 million euros, or 1.37 percent of that of its competitor (see table 11.1). India is the sixteenth most important trading partner for Uzbekistan and the twenty-second for Tajikistan, while China is almost always in the top three with Russia and the European Union. It is the largest trading partner of Kyrgyzstan, the second largest of Uzbekistan and Tajikistan, third for Kazakhstan, and seventh for Turkmenistan, a figure expected to rise in 2010–2011 with the arrival of the first flow of Turkmen gas to Xinjiang.

Several factors explain a gap this large. First, India and Central Asia do not have many institutions of multilateral cooperation that can oversee the development of their relations. The Shanghai Cooperation Organization (SCO) is completely dominated by Moscow and Beijing, India's observer status there has no impact on its economic relations with Central Asia. Second, India and Central Asia have no common borders, while the proximity between Xinjiang,

Table 11.1   Chinese and India Bilateral Trade with Central Asia in 2008

		Kazakhstan	Kyrgyzstan	Uzbekistan	Tajikistan	Turkmenistan
China	Amount in million €	12,237	4,629	1,128	645	519
	Percentage of trade (imports and exports)	18.9	62.3	20.8	20.8	6
	Rank	3rd	1st	2nd	2nd	7th
India	Amount in million €	139	7	59	6.7	36
	Percentage of trade (imports and exports)	0.2	0.1	0.6	0.2	0.3
	Rank	19th	20th	16th	22nd	20th

Source: Table compiled on the basis of official figures available for each country at <http://ec.europa.eu/trade/issues/bilateral/data.htm> (accessed September 23, 2009).

Kazakhstan, and Kyrgyzstan[2] facilitates the trade boom between China and Central Asia. Indian-Central Asian relations are impeded by a multitude of factors resulting less from geographical distance—the two areas are separated only by a few hundred miles, even less through the Wakhan Corridor that links the Tajik Pamir to the northwest regions of Pakistan—and more from an unfavorable geopolitical context stemming from the chronic instability of Afghanistan and Pakistan. Third, the current Central Asian economies need what the Chinese "world's workshop" has to offer: investment in transport infrastructure and energy production, as well as cheap goods that fit in with a low standard of living. Their interest in the Indian "world's back office," with the exception of Kazakhstan, is limited at present. Finally, like Japan, the United States, or the European Union, India is penalized by the private nature of its economy. Indian companies do not receive state support and are particularly constrained in Central Asia by the poor investment climate, while big Chinese firms have the diplomatic and financial clout of Beijing on their side.

The Thirst for Hydrocarbons: The Driving Force of Sino/Indo–Central Asia Economic Relations

Both China and India are driven by their "thirst for energy" and need supplementary sources to sustain their economic growth. They also seek to reduce their dependence on energy importation by tankers, which are much more prone to market instability and geopolitical risks. Thus they covet the continental resources of the Caspian basin (mainly Kazakhstan and Azerbaijan), as well as those, although still poorly estimated, of Turkmenistan and Uzbekistan. In a few years, China has become the second largest consumer of energy in the world, after the United States, and has overtaken Japan as the number two importer of energy in the world, again after the United States. Today China imports more than 40 percent of its energy consumption, mainly from the Persian Gulf and Africa, a figure that could rise to 70 percent by 2020.[3] India imports 70 percent of its crude oil and gas from the Persian Gulf states and could become the fourth largest net importer of oil in the world by 2025. According to the U.S. Department of State, Chinese demand for petroleum will double by 2020, reaching 13 million barrels per day (mbd), while that for natural gas will triple to 100 billion cubic meters (bcm) per year.[4] India's energy demands are expected to reach nearly 3.5 mbd in 2010, while gas consumption will rise to 51 bcm by 2015.[5]

The main players in China's and India's establishment in Central Asia are for China, the China National Petroleum Corporation (CNPC) and its affiliates, such as the China National Offshore Oil Corporation (CNOOC) specializing in foreign investment, the National Oil and Gas Exploration and Development Corporation (CNODC), and the firm Sinopec (China National Petrochemical Corporation); for India, they are the Oil and Natural Gas Company (ONGC), Punj Lloyd, India Oil Corporation (ICO), the national gas company Gail, and Mittal. In Kazakhstan, both Chinese and Indian firms have failed to enter the three main sites, Tengiz, Karachaganak, and Kashagan, which are operated by big Western firms, the state-run corporation KazMunayGas, and, on a smaller scale, by some Japanese and Russian companies. Therefore they must specialize for China in old fields, which are considered technically difficult to exploit, and for India in the petrochemical business. However, their implementation strategies are very different. Chinese firms have Beijing's diplomatic and financial support, enabling them to outbid competitors during negotiations, offer complementary "good neighbor" measures, and accept the authorities' requirement that KazMunayGas be systematically associated with all activities. These strategies elicit angry reactions from competitors, especially Indians, who often perceive Chinese as practicing aggressive energy policy and market distortion.

In less than a decade, Chinese companies have successfully launched themselves into the Kazakh market, and by 2008, they were managing about one quarter of Kazakh production.[6] The general Chinese strategy is to connect all the acquired fields with the giant Sino-Kazakh pipeline, presently under construction. The first section, which became operational in 2003, connects the Kenkiyak field to Atyrau; the second connects the pumping station and railway terminal in Atasu in the Karaganda region to the Dostyk-Alashankou station and was opened in 2006. The third and last section, linking Kenkiyak to Kumkol via the town of Aralsk, is to be completed in 2011. China will thus have the advantage of an oil pipeline more than 3,000-kilometer long connecting the shores of the Caspian to the Dostyk-Alashankou border post, and with a capacity of

20 million tons a year. CNPC's main Kazakh acquisition remains the company AktobeMunayGaz, in which it purchased a 60 percent share in 1997 and a further 25 percent share in 2003. Located in the Aktobe region, it controls almost 15 percent of Kazakh petroleum, and in particular holds a 20-year license for the exploitation of the Zhanazhol and Kenkiyak petroleum and gas sites, which represent 5 percent of Kazakhstan's total petroleum reserves.[7] The offshore Darkhan site is operated by China, which is also involved in more isolated fields that have the advantage of being located along the route of the Sino-Kazakh pipeline (North Buzachi, North Kumkol, and Karazhanbas). CNPC's last acquisition, MangistauMunayGas, which includes the Kalamkas and Zhetybay fields, but not the Pavlodar refinery, a subsidiary of Central Asia Petroleum, was negotiated on an equal basis with KazMunayGas in spring 2009.[8]

Second, China is interested in the gas deposits in Uzbekistan and Turkmenistan. In spite of the challenging regional situation, Beijing has succeeded in convincing Ashgabat, Tashkent, and Astana of building a shared pipeline and jointly selling gas resources. This pipeline, which began to operate in December 2009, will deliver 30 bcm of gas, with expectations of around 50 bcm in a few years.[9] The pipeline starts at the Samandepe well, located near Bagtiyarlyk, on the right bank of the Amu-Darya. It stretches 180 kilometers on Turkmen soil before crossing the Uzbek border at Gedaim. Then it extends for more than 500 kilometers across Uzbekistan, and for nearly 1,300 kilometers across Kazakhstan, before reaching Xinjiang via Shymkent and Khorgos. Kazakhstan, Uzbekistan, and Turkmenistan supply 10 bcm each, but Ashgabat's portion should quickly reach 30 bcm, according to a Sino-Turkmen agreement negotiated in 2006. The CNPC is also the first foreign gas company in Turkmenistan to gain the right to carry out onshore gas extraction activities in the Amu-Darya basin on a production sharing agreement (PSA) basis.[10] Beijing has granted the Turkmen authorities a US$3 billion loan to develop the promising South Yolotan gas field.

India has signed several memoranda of understanding and agreements with Afghanistan, Turkmenistan, Iran, and Kazakhstan on transportation, pipelines, and energy. However, its involvement in Caspian energy is still very modest.[11] Since 2000, Indian companies have tried to get involved in purchasing deposits but have practically never succeeded in obtaining favors from Astana. In 2005, despite an MoU signed between India and Kazakhstan for cooperation in the oil and gas sector, ONGC Videsh, the subsidiary of ONGC that looks after foreign purchases, suffered one of its greatest failures, outbid by the Chinese CNPC for the acquisition of PetroKazakhstan.[12] ONGC Videsh then set its sights on the offshore deposits of Kurmangazy, and Darkhan, but lost the tenders. In 2009, after several additional years of discussions, ONGC-Mittal Energy (OMEL) eventually signed an agreement for joint exploitation of the Satpayev offshore block in the northern Caspian Sea with estimated reserves of

1.85 billion barrels, but the project still needs to be finalized. Mittal Investment Sarl has decided that it would not participate in the Satpayev exploitation but would continue to look for other investment opportunities. As a consequence, ONGC Videsh will be the sole partner of KazMunayGas at the Satpayev site, with a participating interest of 25 percent.[13]

Indian companies should thus be satisfied with ancillary activities in the petrochemical sector. Punj Lloyd has been involved in several engineering and petrochemical infrastructure projects in Kazakhstan since 2002. Its subsidiary Punj Lloyd Kazakhstan (PLK), which has offices in Almaty, Atyrau, and Tengiz, has won many tender bids in this domain, mainly for building crude oil pipelines for AGIP KCO, TengizChevroil, and PetroKazakhstan.[14] Gail, which works with Reliance Industries to jointly set up a mega gas-based petrochemical plant overseas,[15] and Indian Oil Corporation (IOC) have offered to construct a gas treatment station and a refinery at Atyrau and Aktau, making it possible to improve the recuperation of oil from old exploitation fields. Kazakhstan is the second country after Qatar to invite Gail and IOC for talks on the US$ 1.3 billion plant. However, the dossier has not yet been finalized.

Indian interest in Uzbekistan and Turkmenistan is more limited. In 2006, during a visit by Indian prime minister Manmohan Singh to Tashkent, a series of documents were signed between the two countries, including a memorandum between Gail and Uzbekneftegaz. It provides for the joint exploration and exploitation of Uzbek sites, as well as for the construction of liquid gas and oil factories in the western regions of Uzbekistan.[16] In Turkmenistan, ONGC-Mittal Energy made an acquisition in 2007 of 30 percent of the shares of two oil sites (blocks 11 and 12) in the Turkmen sector of the Caspian Sea. It intended to exploit them in partnership with the Danish group Maersk Oil, and the German company Wintershall, a subsidiary of the chemical consortium BASF. In April 2008, a new memorandum between India and Turkmenistan for cooperation in the oil and gas sector was signed. However, in January 2010, ONGC-Mittal Energy exited the blocks after exploratory failures.[17] In Kyrgyzstan, India's Jagson Oil invested more than US$ 1 million to build six fill-up stations in the Osh region.[18]

In spite of these various agreements, India is not yet one of the first ten countries involved in the exploitation of oil and gas resources in Central Asia. It will have difficulties finding a place on this list considering both the already established involvement of Russian and Western companies and the rapidity of China's growth in it. India has placed all its hopes on two gas pipeline projects, the Iran-Pakistan-India (IPI) one, which could run for 1,100 kilometers from the giant South Pars gas field on the Persian Gulf to Gujarat, and in the Turkmenistan-Afghanistan-Pakistan-India (TAPI) one. The first is blocked by the U.S. embargo against Iran; the second is blocked by instability in Afghanistan, Indo-Pak tensions, and price disagreements with Turkmenistan but enjoys strong support from the United States. The American desire to marginalize Russia and Iran and not to promote China in Central Asia opens great opportunities for New Delhi, supported by Washington as an Asian balance vis-à-vis an emergent powerful China, but these expectations have not materialized yet.[19] The shift of oil and gas pipelines to the east is taking place, but not to the south.

Transport Issues and Electricity Exports

India, like China, regards Central Asia as a place of regional crossroads, opening prospects for transit toward Russia, Iran, the Mediterranean, and Europe. Again, the possibilities are numerous, but difficult to achieve. The instability in Afghanistan has so far hindered envisaged corridor or electrical projects, and the poor relations between Central Asian states and their difficulties in agreeing to promote a free flow of goods slowed international ambitions. China is again favored over India as it can benefit from shared borders, especially with Kazakhstan and Kyrgyzstan, while India must pass either through Iran, Afghanistan, the Jammu and Kashmir region, or the difficult Leh-Kashgar pass.

Chinese presence is very important in the infrastructure sector. Beijing and Astana have actively collaborated to modernize their two main passage routes, Dostyk-Alashankou, which is mostly for railroad freight, and Khorgos, for road freight, and to expedite clearance procedures and customs, transforming Kazakhstan into a real "highway" for transcontinental Chinese goods bound for Europe via Russia. In Kyrgyzstan and Tajikistan, Beijing is implementing a two-pronged strategy: first, to improve the border-bound routes in order to increase cross-border transactions, and, second, to open up the most isolated regions in order to facilitate internal communication. Chinese companies are having a noticeable impact in the road sector. They are currently restoring the road from the Irkeshtam border point to the large town of Osh, as well as a section of the Osh-Dushanbe road. They are constructing two tunnels in Tajikistan, namely the Char-Char Tunnel between Dushanbe and Kuliab, and the Shakhristan Tunnel on the road connecting the Tajik capital to Khudjand.[20] In addition, Turkmenistan, Kazakhstan, and Uzbekistan are buying more and more railway equipment from China, including locomotives, passenger wagons, and goods wagons. Despite its isolationist policy, Uzbekistan has also tried to take its place in China's transport dynamics through a railway project linking Xinjiang to the Ferghana Valley.[21]

Beijing is also interested in the Central Asian electricity sector. Contrary to its hydrocarbon policy, the aim of China is not primarily to have this hydroelectricity delivered to the large cities in the east (the electrical lines required would need to stretch over at least 6,000 kilometers), but rather to make up for the energy shortfall in Xinjiang. China would also like to be able to sell Central

Asian hydroelectricity to countries of the southern corridor (Afghanistan, Pakistan, Iran, and India) because of the significant transit fees it would generate. The establishment of Chinese companies in the region, like those it has set up in Russia and Mongolia, had thus been centered around two axes: the construction of new hydroelectric stations and the installation of new electricity lines, in particular high-voltage ones. Astana and Beijing are currently discussing the construction of an electrical coal power station and an 800 kW high-voltage line near the city of Ekibastuz, the total electric output of which would be destined for Xinjiang. In Tajikistan, the Chinese company Sinohydro Corporation is constructing the Zarafshan station near Pendjikent, but Uzbekistan's opposition has stalled the project for the time being. It is also constructing several electric lines in the south heading toward Afghanistan. In Kyrgyzstan, a series of hydroelectric stations has been planned in the Tian-Shan mountains on the border with Xinjiang. Negotiations are currently underway for Chinese financing of the construction of three stations on three cross-border rivers, which would run from the Kyrgyz glaciers toward China.[22]

If the geographical distance between India and Central Asia is not in itself insurmountable, the difficulty of transiting via Afghanistan has blocked the development of trade flows for close to two decades: the absence of secure road, rail, and electric systems prevents practically all north-south transportation. Over the past few years projects for high-voltage line corridors starting in Kazakhstan and Kyrgyzstan and going to Pakistan have been under consideration. In May 2006 in Islamabad, and then in October of the same year in Dushanbe, Tajikistan and Kyrgyzstan confirmed their desire to export energy to Pakistan via Afghanistan. The four countries signed an MoU for the Central Asia-South Asia (CASA) project development, with the support of international donors (the World Bank, International Finance Corporation, Asian Development Bank, Islamic Development Bank, and USAID) and interested private sector investors (the American AES and the Russian RAO-UES). The CASA-1000 project consists of the construction of high-voltage lines between the two grids, which until now have remained without any interconnection. The construction of these lines would give the countries of Central Asia access to the electricity-deficient markets of South Asia for the first time.[23] The Pakistani company NESPAK proposed two possible routes between Tajikistan and Pakistan. The first passes through Kunduz and Kabul (650 kilometers) and the second via the Wakhan Corridor. Although more secure, as it passes only 30 kilometers through Afghanistan, the second route is much more expensive because of the extremely difficult nature of the physical terrain and weather conditions.[24]

However, as with the pipeline projects, these electricity corridors come up against the Afghan question. As an essential transit point for any expansion to the south, Afghanistan suffers from political instability that has largely put the brakes on developing cooperation in electricity with India and Pakistan. The country experiences critical electricity shortages, and public electricity supply there is very limited. Kabul already imports modest quantities of electricity from Turkmenistan, Uzbekistan, and Tajikistan through existing interconnections. The economic and social recovery of Afghanistan will demand a better electrical grid and improved regional integration. The investments made so far in the CASA electricity grid, therefore, provide timely assistance to local populations, although it is more expensive to import electricity from Central Asia than it is from Iran.[25] Mazar-e-Sharif has great potential to become one of the points of distribution for Central Asian energy to Pakistan and India.[26] The CASA-1000 project should advance, at least on the Pakistani side, later in 2010.[27] Several Indian companies (NHPC, BHEL, and PGCIL) have stated their interest in the hydroelectricity sector in Tajikistan and have offered their services for the modernization of the hydroelectric station Varzob I.[28]

On the question of roads, India and Central Asia are members of the International North-South Transport Corridor project, launched in 2000 in Saint Petersburg. Marginalized in the east-west lines supported by the West, India, Iran, and Russia sought to rekindle their complementary economic potential. Their project concentrates on the establishment of a road and rail corridor linking the large cities of Russia with the Iranian ports of the Persian Gulf, and then with Indian Ocean. The first commodity flows began in 2004 but remain limited. India and Pakistan also participate in a joint initiative launched by the Asian Development Bank in 2003, the Central-South Asia Transport and Trade Forum (CSATTF). Several agreements signed between India and Iran, on the one hand, and Tajikistan and Turkmenistan, on the other, have as their objective a reduction of the Astrakhan-Mumbai route by more than 1,000 kilometers in going via Tajik territory and/or via the Caspian Sea, which should enable them to save ten or so days by avoiding the Suez Canal detour. Indian companies are particularly interested in the development of Atyrau and Aktau, which could potentially become ports frequented by Indian products heading for the north.

Geographically, Iran is the best route for Indian exports to and imports from Central Asia, as it allows bypassing Afghanistan.[29] Tehran opened up a transportation route with Turkmenistan in 1996, but here again, flows remain modest. An agreement to build a railway line along the Caspian Sea (Uzen-Gyzylgaya-Bereket-Etrek-Gorgan), signed in 2007 between Russia, Kazakhstan, Turkmenistan, and Iran, will connect Russian railways to the Persian Gulf and will reduce by 600 kilometers the existing connection that goes through Serakhs. The line should be operational by 2011 and allow the transport of about 5.5 million tons in the first year, a figure that is expected to reach 10 million tons annually. It will be supplemented by an Iran-Pakistan rail link between Bafk and Zakhedan. An international highway is also planned along the track. Again, the viability of such a path is not assured. So far, all projects toward Iran have had limited success for reasons relating primarily to the geopolitical status of Tehran and have prevented a connection between the Central Asian and Indo-Pak networks.

Potentially, this North-South Corridor could facilitate the transit of goods from India to Iran's port of Bandar Abbas, and then hopefully to Chabahar. India thus finances the transforming of Chahbahar into a commercial port and its connection with Afghanistan's main ring road highway system (as part of the circular route connecting Herat and Kabul via Mazar-e-Sharif in the north and Kandahar in the south), however the port's capability and its connectivity to the rail networks of Iran and Afghanistan is not up to par.[30] In 2009, the launch of the Northern Distribution Network, charged with supplying nonlethal equipment to international coalition forces in Afghanistan, should speed rail connections between Uzbekistan and Afghanistan and contribute to the launch of a trans-Afghan train, a potential link between Tashkent and Islamabad or New Delhi.[31] For now, Indian access to Central Asia is practical only by air, greatly reducing the profitability of potential exchanges and limiting them to materials with high added value.

The Race for Minerals and Central Asian Uranium

Central Asia has significant reserves of precious and non-precious minerals: gold, uranium, copper, zinc, iron, tungsten, and molybdenum. Kazakhstan is a veritable Eldorado mine, with 8 percent of the world's iron ore. It also boasts the second largest reserves of manganese and the eighth largest reserves of iron and includes nearly one-third of the world's chrome deposits. Coal is not an exception, since the region is ranked as the ninth largest producer. Within the CIS, Kazakhstan has the largest stocks of chromium and lead, ahead of Russia, second in manganese, nickel, silver, and zinc, and third in coal, gold, and tin.[32] Kazakhstan is also second in the world for uranium reserves after Australia, with between 16 and 19 percent of known global reserves, or between 1 and 1.5 million tons. In terms of extraction, in 2007, it was third behind Canada and Australia but hopes to become leading world producer by 2010. The extraction programs, which slowed down in the 1990s, were revived in the following decade: 2,000 tons in 2000, 8,500 in 2008, and approximately 13,000 tons in 2009.[33] The government plans to extract around 15,000 tons in 2010 and hopes to reach 30,000 tons by 2020.

China is becoming increasingly present in the Central Asia mineral industry. Beijing is very interested in Kazakh and Kyrgyz gold. In 2005, the China National Gold Group Association and the metallurgic complex Kazakhaltyn Mining signed a contract for a joint venture to exploit Kazakhstan's gold deposits.[34] In June 2006, China proposed to Bishkek the formation of a

Sino-Kyrgyz joint venture to extract Kyrgyz gold deposits, 10 to 20 tons of which would be held at the Chinese Development Bank as a credit guarantee. Kyrgyzstan, however, rejected the offer.[35] In May 2008, China followed in Russia's steps by establishing itself in the development of aluminum, the main industry of Tajikistan. The Tajik aluminum company TALCO and the Chinese National Corporation for Heavy Machinery (CHMC) signed an agreement for the construction of two factories in the Yavan district that will supply TALCO with raw aluminum for further refinement.[36] Here again, India may come into competition with China. In 2006, the Indian government stated its interest in Kazakhstan's metal resources, in particular for gold as well as precious and semiprecious stones. Several joint ventures in the jewelry sector have been created. New Delhi also wants to invest in Tajik mines, and in the development of cement production, but investment conditions for now impede the finalization of these projects.

The only massive Indian presence in the Central Asian metal extraction industry is Arcelor Mittal, but the company is a multinational holding that has little to do with the interests of New Delhi and is considered Indian by virtue of the nationality of its director, Lakshmi Mittal. He acquired KarMet, the metallurgical combine of Karaganda, in 1995.[37] According to the Indian Foreign Ministry, total investments in this project have reached US$ 2 billion. This industrial giant of Kazakhstan now comprises of six steel mills, coke and steel plants, rolling, melting, several rounds of tubing production and metal coating (for aluminum, zinc, and polymer), and two electricity plants and operates ten mines and a dozen surface mining sites.[38] Tensions between the Kazakh authorities and Arcelor Mittal, which enjoys significant tax privileges negotiated during its implementation, grew further around 2008–2009. The various charges are legal, environmental, and social ones. Several fatal accidents in coal mines, for example, altered the Kazakh public opinion, which accused the company of not upgrading its obsolete Soviet security systems. In 2008, the Kazakh government threatened to withdraw Mittal's license but could not afford such a loss of production and the unemployment of tens of thousands in an already stricken region. Although Mittal Arcelor is a multinational, it plays up its image as Indian in Kazakhstan by financing some Indian cultural activities. This national side is highlighted in the local media, thus leaving the diplomatic mission in New Delhi in a quandary.

China also needs uranium, chiefly to complete the construction of several tens of nuclear power plants. The 2005 strategic cooperation treaty fosters the strengthening of ties between Beijing and Astana in the atomic energy sector and mentions "the unification of more segments of the industrial cycle for the production of enriched uranium."[39] Starting in 2006 and 2007, several cooperation agreements were signed between Kazatomprom and the Guangdong Nuclear Power Group (CGNPC). In 2008, a tripartite strategic partnership between the Kazakh national company and two Chinese state companies, the CGNPC and the China National Nuclear Corporation (CNNC), propelled Kazakhstan to the rank of the largest foreign supplier of uranium to China, surpassing what was its traditional partner, Areva. The two Chinese companies have invested large sums of money in three joint mining ventures.[40] Overall, Kazakhstan is expected to provide about 24,000 tons of uranium to China until 2020. Astana has also set itself up in the Chinese market for building nuclear power plants. Kazakhstan will provide support to Beijing for its VBER-300 reactors and fuel for a range of new plants, some of which may be built outside of China. The Guangdong Nuclear Uranium Corporation has also had a presence in Uzbekistan since 2009, when it signed an agreement with the State Committee for Geology and Mineral Resources. This established a joint venture, Uz-China Uran, to explore the Boztauskoe deposits in the Navoi region, including uranium that will be sold by the Chinese company.[41]

The situation is different for India, which has long been marginalized in the uranium market due to its refusal to accede to the Nonproliferation Treaty. New Delhi signed a memorandum of understanding with Kazakhstan in January 2009. Nuclear sector is now perceived as a major niche of Indo-Kazakh partnership and collaboration is planned for several sections: uranium mining, personnel training, and fuel supply for India's nuclear industry. India indeed lacks the necessary uranium resources to meet its growth in energy demand. Since its agreement with the International Atomic Energy Agency, it is only beginning to participate in this global market. Kazakhstan should provide some unknown amounts of uranium to New Delhi starting in 2010.[42]

Chinese Commercial Impact in Central Asia

China is on its way to becoming the primary economic partner of the Central Asian states. In trade terms, it will compete not with Russia, but with the European Union. Sino-Central Asian trade mostly involves trade between China and Kazakhstan (about 80 percent of the total), more than two-thirds of which is done with Xinjiang. Astana quickly rose to become the second largest of China's trading partners in the CIS after Russia and has for quite some time already held the mantle of Xinjiang's largest foreign trading partner. The Chinese neighborhood has drastically altered the economies of the Central Asian states. Kyrgyzstan has become one of the main places for the re-exportation of Chinese products throughout the rest of Central Asia: about 75 percent of Chinese imports to Kyrgyzstan are re-exported. The extent of this commercial growth is such that the re-exportation of Chinese goods has become one of the two largest economic activities of Kyrgyzstan after gold extraction.[43] This situation reinforces Central Asian economic specialization in raw materials. More than 80 percent of Central Asian exports to China are composed of oil and gas products as well as ferrous and nonferrous metals, while Chinese finished products—textiles, toys, shoes, and electrical and electronic goods—account for about 90 percent of Chinese exports to Central Asia.[44] Although China is primarily concerned with the resources in Central Asia and the transport potential of the region, it also invests in areas such as construction materials (wholesale bazaars specializing in construction are dominated by Chinese products) and telecommunications. One can see a more modest presence in Uzbek textiles and agribusiness and in Kazakh IT projects.

The Chinese presence is, of course, beneficial to the Central Asian economies, but in an ambiguous way, since above all else it privileges the heavy industry sectors, which are in the hands of the oligarchs and power clans. Small- and medium-sized Chinese enterprises are rare, since the Central Asian market is very limited and the investment climate is seen as negative. Trade has given rise to some private enterprises, whether Chinese or Central Asian, or to joint ventures owned by the middle classes; nonetheless, here too business benefits corrupt milieus, customs officers, the police, and other authorities. In addition, Central Asian public opinion increasingly condemns Chinese methods of economic settlement. Chinese firms come with their own equipment and materials and do not give work to local enterprises. The personnel is mostly comprised of Chinese workers who live in isolation at their place of work, without much interaction with the host society, and the few locals employed are often submitted to appalling working conditions. This calls into question whether the Chinese presence brings development with it, and whether it contributes to indigenous know-how and techniques, local training, and interaction with the settlement country. Similar questions are being raised regarding the Chinese involvement in Africa and closer, in Afghanistan, and the response is paradoxical indeed.

The long-term implications of China's engagement in landlocked Central Asia in terms of transit and transport will partially determine the future of the region. Chinese investments in infrastructure will enable the Central Asian states to escape from the increased isolation from which they have suffered following the disappearance of Soviet-era infrastructure networks. They benefit from consumer products that are appropriate to their low standard of living, but which are also capable of satisfying the growing technology consumption needs of the middle classes, in particular in Kazakhstan. The massive influx of Chinese products will give the peoples of Central Asia the opportunity to resume their traditional role as a transit culture exporting goods as far as Russia, something that the Kyrgyz and Uzbek migrants in Russia are already starting to do. This trade also ensures that the area can be turned into a platform for the re-export of Chinese products and offers a new range of jobs in the tertiary sector for Central Asians. If Beijing has not yet managed to develop cultural diplomacy, it has nonetheless undertaken to strengthen its linguistic influence in Central Asia. This has been well received among the younger generations seeking profitable career opportunities. Fluency in Chinese today guarantees an extremely quick rise up of the Central Asian social ladder in both the public administration, especially in the Ministry of Foreign Affairs, and the private sector, especially those relating to trade, transit, freight, legal supervision, and translation.[45]

Future Niches for India: Space, IT, Pharmaceuticals, and Textiles

Unlike China, there are no truly large Indian investments in the Central Asian economy, apart from Mittal's metallurgical project in Kazakhstan. The economic niches for Indian companies are also much more limited. Except mining, electric power, oil, and gas, the most promising areas of cooperation seem to be in the textile industries, silkworm breeding, agriculture, space technology, pharmaceuticals, and, of course, high technology. Moreover, while large Chinese banks are well established in Central Asia through partnerships with local banks, India has only the Punjab National Bank, which recently opened a branch in Kazakhstan.

Since the 1950s, Kazakhstan has hosted on its territory the famous Baikonur Cosmodrome, but it has otherwise been a passive player in Russia's space pursuits. Yet in recent years, Kazakh government authorities, interested in sharpening their country's international image, have sought to become a part of the space race, viewed as a technological challenge that confirms a rise to great power status. Although Moscow will remain Astana's primary partner, New Delhi is also trying to position itself in this promising sector. India's technical skills are indeed particularly welcome. In October 2007, a delegation from Kazakhstan's space agency, led by former cosmonaut Talgat Musabaev, met with Madhavan Nair, the president of the Indian Space Research Organization (ISRO), in order to establish bilateral cooperation in space affairs.[46] The two countries share very similar goals in space, although India is far more advanced, and both seek new partners. They discussed the creation in Kazakhstan of a landing space, the launch of Indian IRS rockets, which are known for the quality of the images they take in space, and the use of a radar complex on Kazakh territory. Astana, for its part, was particularly interested in the technology center in Bangalore that houses the ISRO Rocket Center. New Delhi has already made a name for itself on the world market for launching heavy satellites and could help Kazakhstan to attract non-Russian projects.[47]

Space cooperation between Kazakhstan and India is expected to grow in the coming years. It remains to be seen whether this cooperation will move forward at Russia's expense or with its consent—the latter, however, seems more plausible. Russia cannot advance in this sector without Baikonur, while New Delhi needs to share technology with Russia, and Astana does not have the training facilities and technical expertise needed to independently manage its space ambitions. The Russian-Indian rapprochement, especially in the military-industrial sector, could give birth to triangular space relations between Astana, Moscow, and New Delhi. Such a situation would not be without self-interest and would quietly put aside Chinese cooperation that tends to come with heavy consequences.[48]

New Delhi also hopes to gain access to Central Asia through arms sales and technology transfers, which are relatively easy because they can be sent by air and do not require transit by road or train through Afghanistan. Offering the possibility of significant technology transfer to Central Asia, India hopes to mitigate its weak military presence in the region by enhancing its image as a new technological power.

Indian know-how is, of course, particularly famous in the information technology sector. The Central Asian states are on a quest for development in this domain, especially Kazakhstan, which launched a Program of Non-industrial Development and Investment for the period of 2003 to 2015. During Nursultan Nazarbaev's 2002 visit to India, this sector was decreed one of the principal areas for bilateral cooperation. In 2001, the state educational center Bilim and the Chennai-based Internet Business Factory India (IBFI) opened a technopole devoted to the handling of information technology in the Kazakh public school system. IBFI has also offered to set up an intranet system for the Kazakh national education system.[49] In 2005 the first Kazakh-Indian information technology center was opened at al-Farabi University. The umbrella organization for all IT companies of India, the National Association of Software and Service Companies (NASSCOM), is actively involved in Indian-Kazakh cooperation in the IT sector. The company STPI Bangalore has been very active in the venture to open Kazakhstan's first information technology park in Alatau, near the former capital Almaty.[50] India has also formed centers of information technology in Tashkent and Dushanbe. In 2006, memoranda of intent were signed to open such centers in Ashgabat and Bishkek. In 2004, New Delhi gave to Uzbekistan a grant of US$ 650,000 to open an Uzbek-Indian information technology center at the National University of Tashkent. Since the beginning of the 1990s, an Indian Technical Economic Cooperation Program (ITEC) has been operational in Central Asia. More than 600 specialists from Uzbekistan, 561 from Kazakhstan, 200 from Turkmenistan, and 343 from Tajikistan have been trained under this program.[51]

India is also one of the main exporters of pharmaceutical products to Central Asia. By contrast to Russia, known for its bad imitations, it enjoys a solid reputation in this area. Currently, New Delhi supplies Central Asia with more than

30 percent of its pharmaceutical needs.[52] Several joint ventures that specialize in the manufacture of pharmaceutical products have been created, such as Gufic Avicenna, Shreya Life Scientist Pharmaceutical, and Reddy-Pharmed. Kazakhstan represents a particularly interesting market in this domain, with an estimated worth of US$ 400 million per year, and an availability of 6,000 preparatory products of which only 10 percent are produced locally.[53] Several large Indian pharmaceutical companies have thus opened offices in Almaty or Astana. Tajikistan, particularly needy in this sector, is also the object of Indian export policies. In Turkmenistan, India financed the creation in 1998 of a joint venture called Turkmenderman Adjanta Pharma, which today offers close to seventy pharmaceutical preparations.[54] Finally, thanks to its long tradition in the textile domain, India is particularly sought after in Central Asia, above all in Uzbekistan, whose government is seeking to reduce the rather unprofitable export of raw cotton and to increase the local production of finished products.[55]

Conclusion

In economic terms, the presence of China and India in Central Asia really differs. Beijing is a global power, increasingly present in all sectors, whether by its imports, its exports, or its ability to offer cheap technology, while India is a minor economic player with specialized niches, but lacking total reach. As such, the economies of India and Central Asia are complementary. However, basic Indian products cannot compete with Chinese ones and face major transit problems. In this growing competition with China, India remains subject to a geopolitical context over which it has very little control: the stabilization of Afghanistan, the regional integration of Iran, the deteriorating situation in Pakistan, and the speed of Chinese settlement.[56] Furthermore, areas where India can compete with China, such as knowledge technologies, are still relatively underdeveloped in Central Asia. They mainly concern Kazakhstan and affect only the middle and upper classes, a small portion of the population. However, these sectors are expected to grow and so are the implementation prospects of Indian companies: although for now Central Asia needs the Chinese "world's workshop," in the future it will have to focus on the Indian "world's back office." Moreover, in the decades ahead, the fear of a near-total domination of China in the region should push Central Asian political elites to give special attention to other actors. India will then find more reasons to look beyond Afghanistan and will probably assert itself more clearly, in economic terms, in the region.

Notes

1. Senior Research Fellow, Central Asia-Caucasus Institute and Silk Road Studies Program, SAIS, Johns Hopkins University, Washington, DC.

2. Despite its shared border with Xinjiang, Tajikistan is less advantaged in geographical terms.

3. Sh. Shuja, "China's Energy Needs and Central Asia," *National Observer*, Council for the National Interest, Melbourne, 67 (2006), pp. 56–65.

4. D. L. O. Hayward, "China's Oil Supply Dependence," *The Journal of Energy Security*, June 18, 2009.

5. Planning Commission; Government of India, Tenth Five-Year Plan (2002–2007), quoted in V. Kumar Bhatia, "India-Kazakhstan Relations: Challenges and Prospects," *Mainstream*, vol. XLVII, no. 38, September 5, 2009.

6. But this proportional share of total Kazakh oil production will decrease in the forthcoming years as the exploitation of Tengiz and Kashagan increases the size of the pie to China's disadvantage. See S. Peyrouse, "Chinese Economic Presence in Kazakhstan: China's Resolve and Central Asia's Apprehension," *China Perspectives*, no. 3, 2008, pp. 55–75.

7. "Prisutstvie Kitaia v Kazakhstane: 'Dostizheniia' i 'perspektivy,'" *Zona.kz*, <http://zonakz.net/articles/?artid=13774> (accessed August 17, 2008).

8. "CNPC and KazMunayGas Acquires 100 percent of MangistauMunaiGas Shares," CNPC website, April 24, 2009, <http://www.cnpc.com.cn/en/press/newsreleases/CNPCandKazMunayGasacquires100ofMangistauMunaiGasshares.htm> (accessed January 18, 2010).

9. "Turkmenistan: Gas Pipeline to China Is Ready," *Asia News*, August 22, 2006, <http://www.asianews.it/index.php?l=en&art=6997> (accessed August 17, 2008).

10. J. Šír, S. Horák, "China as an Emerging Superpower in Central Asia: The View from Ashkhabad," *The China and Eurasia Forum Quarterly*, vol. 6, no. 2, 2008, pp. 75–88.

11. S. Moore, "Peril and Promise: A Survey of India's Strategic Relationship with Central Asia," *Central Asian Survey*, vol. 26, no. 2, 2007, pp. 279–291.

12. R. Dwivedi, "China's Central Asia Policy in Recent Times," *The China and Eurasia Forum Quarterly*, vol. 4, no. 4, 2006, pp. 139–159.

13. "Note on Satpayev Exploration Block, Kazakhstan," ONGC Website, November 18, 2009 <http://www.ongcindia.com/press_release1_new.asp?fold=press&file=press417.txt> (accessed January 18, 2010); "Mittal exits Kazakh oil block exploration," *Thaindia*, <http://www.thaindian.com/newsportal/business/mittal-exits-kazakh-oil-block-exploration_100276700.html#ixzz0fQNF8r62> (accessed January 18, 2010).

14. See Punj Lloyd Kazakhstan website, <http://www.punjlloyd.com/subpage.php?opt=&page_cat=2&id=15> (accessed January 18, 2010).

15. "Kazakhstan Wants GAIL, IOC to Set Up Petrochem Plant," *Press Trust of India*, January 27, 2009, <http://www.business-standard.com/india/news/kazakhstan-wants-gail-ioc-to-setpetrochem-plant/347179/> (accessed January 18, 2010).

16. "Gail to Set up LPG Plants in Uzbekistan," *Business Standard India*, May 8, 2006, <http://www.business-standard.com/india/news/gail-to-setlpg-plants-in-uzbekistan/410/on> (accessed January 18, 2010).

17. "ONGC-Mittal Exit Turkmenistan Oil Block," *The Indian Time*, January 20, 2010.

18. Irina Komissina, "Will India Become a Full-fledged Participant in the Big Game in Central Asia?" *Central Asia and the Caucasus*, no. 1, 2008.

19. See F. Starr (ed.), *The New Silk Roads: Transport and Trade in Greater Central Asia* (Washington: Central Asia-Caucasus Institute, 2007).

20. For more details, see S. Peyrouse, "The Economic Aspects of the Chinese-Central-Asia Rapprochement," *Silk Road Papers* (Washington: Central Asia and Caucasus Institute), September 2007.

21. S. Peyrouse, "The Growing Trade Stakes of the Chinese-Kyrgyz-Uzbek Railway Project," *Central Asia and Caucasus Analyst*, March 11, 2009.

22. S. Peyrouse, "The Hydroelectric Sector in Central Asia and the Growing Role of China," *The China and Eurasia Forum Quarterly*, vol. 5, no. 2, 2007, pp. 131–148.

23. V. Vucetic and V. Krishnaswamy, *Development of Electricity Trade in Central Asia–South Asia Region* (Washington: World Bank, 2005).

24. Ibid., p. 12.

25. The average tariff/kWh varied from about 2.5 cents in the Herat area to 6.9 cents to 7.4 cents in the Kunduz and Mazar-e-Sharif areas. Ibid, p. 19.

26. World Bank, *Central Asia Regional Electricity Export Potential Study* (Washington: World Bank, December 2004), p. 35.

27. "Work on CASA 1000MW Import Project to Start in '10," *The Nation*, July 22, 2009.

28. "India Launches Power Station Upgrade for Tajikistan's Varzob-1," *AsiaPulse News*, September 1, 2008. See also the NHPC Website, <http://www.nhpc.co.in> (accessed January 18, 2010).

29. G. Sachdeva, "India," in F. Starr (ed.), *The New Silk Roads: Transport and Trade in Greater Central Asia* (Washington: Central Asia-Caucasus Institute, 2007), pp. 335–382.

30. E. Koolaee, M. Imani Kalesar, "India's Energy Security toward the Caspian Sea Region: A Critical Review," *The China and Eurasia Forum Quarterly*, vol. 8, no. 1, 2010, pp. 83–94.

31. Kuchins, T. Sanderson, D. Gordon, *The Northern Distribution Network and the Modern Silk Road* (Washington: CSIS, December 2009).

32. R. M. Levine, G. J. Wallace, "The Mineral Industries of the Commonwealth of Independent States," in *2005 Mineral yearbook* (U.S. Geological Survey, 2007).

33. "Progress in Kazakh Ambitions," *World Nuclear News*, August 12, 2009, <http://www.world-nuclear-news.org/ENF-Progress_in_Kazakh_ambitions-1208095.html> (accessed November 4, 2009); "Kazakhstan vykhodit na pervoe mesto v mire po dobyche urana," *RIA Novosti*, December 15, 2009.

34. "Kitaiskaia i kazakhstanskaia kompanii sozdali SP po dobyche zolota v Kazakhstane," *RosInvest*, September 7, 2005, <http://www.rosinvest.com/news/126039/> (accessed June 11, 2007).

35. "Kitai gotov investirovat' 900 millionov dollarov v Kyrgyzstane v obmen na resursy i stroitel'stvo zh.d.," *CentrAsia*, January 29, 2005, <http://www.centrasia.ru/newsA.php4?st=1106979480> (accessed June 11, 2007).

36. S. Peyrouse, "China's Recent Advance in Central Asia," *Central Asia and Caucasus Analyst*, December 12, 2008.

37. S. Smirnov, "Kazakh Metallurgy under the Control of International Corporations," *Kazakhstan International Business Magazine*, no. 2, 2008, <http://www.investkz.com/en/journals/56/467.html> (accessed November 4, 2009).

38. See Arcelor Mittal Temirtau website, <http://arcelormittal.kz/about/o_kompanii/> (accessed November 6, 2009).

39. S. Shmidke, "Atomnaia promyshlennost' Kazakhstana: Sovremennoe sostoianie i perspektivy razvitiia," PIR center, 2006, <http://www.pircenter.org/data/publications/06–05–31Shmidke_Article.pdf>.

40. "Press Release on Signing of Agreements for Strategic Partnership between Kazatomprom, CNNC and CGNPC," Kazatomprom website, November 06, 2008, <http://www.kazatomprom.kz/en/news/2/Press_Release_on_Signing_of_Agreements_for_Strategic_Partnership_Between_Kazatomprom_CNNC_and_CGNPC> (accessed November 6, 2009).

41. "Uzbekistan i Kitai sozdaiut uranovoe SP," *Gazeta.ru*, August 31, 2008, <http://www.gazeta.uz/2009/08/31/uranium/> (accessed September 22, 2009).

42. "Kazakhstan and India Sign Nuclear Cooperation Accord," *World Nuclear News*, January 26, 2009, <http://www.world-nuclear-news.org/newsarticle.aspx?id=24507> (accessed November 6, 2009).

43. B. Kaminski, G. Raballand, "Entrepôt for Chinese Consumer Goods in Central Asia: The Puzzle of Re-exports through Kyrgyz Bazaars," *Eurasian Geography and Economics*, 2009, vol. 50, no. 5, pp. 581–590.

44. V. Paramonov, A. Strokov, *Ekonomicheskoe prisutstvie Rossii i Kitaia v Tsentral'noi Azii* (Central Asian Series, Conflict Studies Research Center, Defence Academy of the United Kingdom, Shrivenham, 07/12, 2007), p. 5.

45. M. Laruelle, S. Peyrouse, *China as a Neighbor. Central Asian Perspectives and Strategies* (Washington: Central Asia-Caucasus Institute, 2009).

46. "Kazakhstan for Tie-up with ISRO," *The Hindu*, September 30, 2007, <http://www.thehindu.com/2007/09/30/stories/2007093055171400.htm>> (accessed January 12, 2010).

47. "Kazakhstan Indefinitely Postpones Space Program," *China View*, April 10, 2009, <http://news.xinhuanet.com/english/2009–04/10/content_11162590.htm> (accessed January 12, 2010).

48. S. Peyrouse, "Russia and India Face Kazakhstan's Space Ambitions," *Central Asia and Caucasus Analyst*, October 29, 2008.

49. P. Sneha Chrysolite, "IBFI and Republic of Kazakhstan Enter into Education JV," *IT People*, December 3, 2001, <http://www.itpeopleindia.com/20011203/careers1.shtml> (accessed January 14, 2010).

50. "India's Substantial Participation in IT Park, Alatau," Indian Embassy in Almaty Website, September 15, 2006, <http://www.indembassy.kz/150906.html> (accessed January 15, 2010).

51. Komissina, "Will India Become a Full-fledged Participant?"

52. Ibid.

53. Sathaye, "India Kazakhstan Sign 5 Pacts to Bolster Partnership," *Sarkaritel*, January 27, 2008, <http://www.sarkaritel.com/news_and_features/january2009/27indkazakhstanpacts.htm> (accessed January 12, 2010).

54. "Pervye shagi," *Farmacevticheskii vestnik*, no. 3, 1999, <http://www.old.pharmvestnik.ru/cgi-bin/statya.pl?sid=1240> (accessed June 26, 2008).

55. As such, in 2006, Spentex Industries, one of the leading companies in the Indian textile market, acquired the national company TTTL in Tashkent (Tashkent-To'yetpa Tekstil Limited). This happened to be the first direct investment by an Indian company in Uzbekistan. Spentex also acquired capital from Farg'ona IpGazmala Ishlab Chiqarish, the Margelan textile factory in the Ferghana Valley, which specializes in the dyeing of work clothes and industrial furniture and has a reported dyeing capacity of 30,000 meters per day.

56. G. Sachdeva, "India's Attitude towards China's Growing Influence in Central Asia," *The China and Eurasia Forum Quarterly*, vol. 4, no. 3, 2006, pp. 23–34.

CHAPTER 12

The Reconstruction in Afghanistan: The Indian and Chinese Contribution

GULSHAN SACHDEVA[1]

Afghanistan has witnessed diverse projects of nation building and sociopolitical transformation in the recent decades. The Soviet project of building communism in Afghanistan resulted in over 1 million dead and 5 million Afghan refugees, mainly in the neighborhood. Similarly, when Pakistan pushed the conservative Taliban regime in Afghanistan, the world faced disastrous consequences, including 9/11. The current international project of building democracy and market economy is mandated by the United Nations and being implemented mainly by the Western alliance led by the United States. So far, this endeavor has produced mixed results. Apart from installing a democratic government, the country has made significant achievements in infrastructure, education, and the economy in the last eight years. After reaching record levels in 2007, opium cultivation and production have somewhat stabilized at moderate levels in the last two years. Although the alliance has had significant successes in many areas, the Taliban insurgency is gaining strength in some parts of the country and security situation has deteriorated. There is also an alarming rise in suicide bombings. Most analysts believe that there is a need to rethink the present strategy.[2]

The new Af-Pak policy in Afghanistan has failed to show any significant improvement. Instead of weakening, antigovernment forces have been able to increase its strength even in northern Afghanistan.[3] Within six months of announcing a "comprehensive new strategy for Afghanistan and Pakistan" in which the new U.S. president Barack Obama aimed "to disrupt, dismantle and defeat al Qaeda in Pakistan and Afghanistan, and to prevent their return to either country in the future,"[4] he was again considering shifting his strategy.[5]

The new Afghanistan-Pakistan Regional Stabilization Strategy outlined by the U.S. State department in January 2010 focuses on reintegration, expanded civilian presence, and regional diplomacy. At the recent London Conference participants "re-affirmed the goals of greater Afghan leadership, increased regional cooperation and more effective international partnership." To end the stalemate, consensus is also emerging on reconciliation with the Taliban.

With continuing excessive focus on security, narcotics, and corruption in the Western media, relatively less attention has been paid to India and China in Afghan reconstruction as well as developments in the area of regional cooperation. This chapter also argues that despite difficult security situation and limited capacities, Afghanistan could emerge as an important player in regional economic cooperation. All international and regional players have appreciated its approach toward regional cooperation. High economic growth in both Central and South Asian regions is also pushing policymakers to work for integration strategies. It is further argued that developments in the area of regional cooperation involving Afghanistan have major implications for regional peace and stability as well as India's linkages with the Eurasian region.

Background

Decades of war, followed by the Taliban regime, destroyed the core institutions of Afghan state. The heavily war-torn economy faced high levels of absolute poverty, ill health, large-scale illiteracy, and complete marginalization of female population. In addition, millions of Afghans left the country and became refugees mainly in the neighboring countries. After the fall of Taliban, all Afghan factions who were opposed to Taliban met in Bonn in December 2001. The meeting was sponsored by the United Nations. The Bonn Agreement[6] charted the roadmap for the political transformation of the country into a democratic state. The UN Security Council endorsed the agreement through its resolution 1383.[7] Under the leadership of Hamid Karzai, a transitional administration was established to guide the process of transformation. The interim administration derived its authority through the *Loya Jirgha* (Grand Council) convened by former Afghan king Zahir Shah. In early 2002, international donors pledged about US$4.5 billion in Tokyo for the reconstruction efforts in Afghanistan. In March 2002, the UN also established the UN Assistance Mission in Afghanistan (UNAMA). The constitutional *Loya Jirgha* adopted a new constitution in January 2004, with the presidential form of government. In April 2004 in Berlin, 23 donor nations pledged a total of US$8.2 billion in aid to Afghanistan over three years.[8]

Under the new constitution, presidential elections were held in October 2004 and parliamentary elections in September 2005. More than 75 percent voters participated in the presidential election and a significant number of women were elected to the National Assembly. The constitution also established legal protection for private property and it was presumed that economic development in the country will be based mainly on the market economy. These were remarkable achievements for a country destroyed by decades of war. Once these landmarks were achieved, international community and the Afghanistan government agreed on the *Afghanistan Compact*[9] at the London Conference in 2005. The compact set ambitious targets for security, governance, development, regional cooperation, and counter-narcotics. While reviewing the *Afghanistan Compact,* at another international conference on Afghanistan in Paris in June 2008, the international community made further commitments for the next five years.[10] Despite serious difficulties, the process of the second presidential election was completed in 2009.

Achievements

According to the *Afghanistan National Development Strategy*[11] (ANDS) more than 5 million Afghan refugees have returned home since 2002. In 2006 alone 342,925 Afghan refugees returned from Pakistan and Iran and another 1,000 from other countries. The number of school-going children has grown from under 1 million in 2001 to about 6 million in 2007 (one-third of them are girls). In 2007, there were more than 9,000 schools (including 1,337 all-girls and 4,325 co-educational). The number of teachers has increased sevenfold to 142,500, which included 40,000 female teachers. In the health and nutrition sector, an amount of more than US$1 billion has been invested in the last five years. As a result, percentage of people living in areas where basic health care facilities are made available has increased from 9 percent in 2002 to 85 percent in 2008. Infant mortality rate has been reduced by 26 percent in five years[12]; 76 percent of children under the age of five have been immunized against childhood diseases. Between 2002 and 2008, there has been a 38 percent reduction in child mortality and 40 percent in maternal mortality. As a result, the lives of approximately 500,000 children have been saved since 2003.[13]

It is remarkable that despite a difficult legacy, the macroeconomic stability in Afghanistan has been maintained in the last few years (see table 12.1). This has been a result of disciplined fiscal and monetary policies. A new currency was successfully introduced. Till 2007, inflation remained reasonably low and exchange rate has been stable. More than a dozen private commercial banks, two private airlines, and 13 microfinance institutions are operating successfully. About 150 cities across Afghanistan now have access to mobile phone

Table 12.1 Some Macroeconomic Indicators in Afghanistan

Year	GDP Growth	Inflation	Exchange rate to US$ (annual average)
2004	8.0	N.A	47.8
2005	16.1	12.3	49.7
2006	8.2	5.3	49.9
2007	12.1	12.9	49.8
2008	3.4	28.3	50.2
2009*	9.0	6.0	N.A
2010*	7.5	6.8	N.A

Note: *projected
Source: Asian Development Outlook 2009 (ADB: 2009), pp. 296, 303, 314.

networks and Internet provider services. Many multinationals are already operating or showing an interest in Afghanistan, which include Coca Cola, Siemens, Nestle, and Etisalat. In 2001, only 15,000 people had access to telecom facilities. Today the number of telephone users (mainly mobile) has crossed 4 million, which is more than 15 percent of the population. Afghan Telecom has installed 86,000 fixed digital lines and 233,000 wireless lines in all 34 provinces. By mid-2009 the process of constructing a 3,200-kilometer optical-fiber network connecting major provincial capitals with one another and also with neighboring countries will be complete. A largely free and privately owned media sector has developed. Presently Afghanistan has seven national TV stations (out of which six are private), numerous radio networks, and a diverse and increasingly professional print media. According to government sources, about 12,000 kilometers of roadways have been rehabilitated, improved, or built, including the 2,200-kilometer-long ring road that connects all major towns of Afghanistan, national highways, provincial roads, and rural roads. More than US$2 billion has been spent on roads. All these projects are implemented in difficult security situations that are normally not mentioned while discussing broader security and strategic matters concerning Afghanistan.

Security Situation and Narcotics

After initial successes till 2004–2005, situation in Afghanistan has become more difficult, complex, and challenging. One of the main reasons has been the deteriorating security situation, particularly in the south and east of the country. A major change that has happened in the last three years is the rise in suicide bombings that reached almost 150 in 2007 and further increased in

Table 12.2   Coalition Military Fatalities in Afghanistan, 2001–2009 (by year and month)

Year	Jan	Feb	Mar	Apr	May	Jun	Jul	Aug	Sep	Oct	Nov	Dec	Total
2001	0	0	0	0	0	0	0	0	0	3	5	4	12
2002	10	12	14	10	1	3	0	3	1	6	1	8	69
2003	4	7	12	2	2	7	2	4	2	6	8	1	57
2004	11	2	3	3	9	5	2	4	4	8	7	1	59
2005	2	3	6	19	4	29	2	33	12	10	7	4	131
2006	1	17	13	5	17	22	19	29	38	17	9	4	191
2007	2	18	10	20	25	24	29	34	24	15	22	9	232
2008	14	7	19	14	23	46	30	46	37	19	12	27	294
2009	25	24	28	14	27	38	76	77	70	74	32	35	520

Source: http://icasualties.org/oef/ (accessed on 15 January 2010)

2008. As table 12.2 shows the figures of coalition casualties in Afghanistan are growing with every passing year making 2009 the bloodiest year since 2001.

In addition, opium production continues to be a serious problem in Afghanistan although 2008 and 2009 opium surveys by the United Nations Office of Drugs and Crime (UNODC) have shown that there has been some moderate decline. In 2008 it was shown that 98 percent of the total cultivation was confined to seven provinces with serious security problems. Five out of these provinces were in the south and two in the west of Afghanistan. The seven provinces that contributed to 98 percent of Afghan opium cultivation and production in 2008 were Hilmand, Kandahar, Uruzgan, Daykundi, Zabul, Farah, and Nimroz, clearly showing strong linkages between local opium production and the security situation. Opium cultivation in Afghanistan has decreased by 22 percent, from 157,000 hectares in 2008 to 123,000 hectares in 2009. In 2008, Hilmand remained the single largest opium-cultivating province where 66 percent of the total Afghan opium cultivation was done. In 2009 cultivation declined by a third, to less than 70,000 hectares in Hilmand. The 2009 survey also shows that compared to 13 in 2007 and 18 in 2008, 20 provinces (out of 34) were poppy free. All the seven provinces in the northern region have been poppy free for almost a decade. The survey rightly asserts that "controlling drugs in Afghanistan will not solve all of the country's problems, but the country's problems cannot be solved without controlling drugs."[14]

Various official and unofficial reports and studies have broadly pointed out that Afghanistan today is at crossroads. The *Afghanistan Study Group* report that was released in early 2008 sums up the mood by asserting that "the progress achieved after six years of international engagement is under serious threat from resurgent violence, weakening international resolve, mounting regional challenges and a growing lack of confidence on the part of the Afghan people about the future direction of their country"[15]

Similarly, the *Centre for Strategic and International Studies* (CSIS) report[16] sums up its findings as follows: (1) Afghans are losing trust in their government because of an escalation in violence; (2) Public expectations are neither being met nor managed; and (3) Conditions in Afghanistan have deteriorated in all key areas targeted for development, except for the economy and women's rights. In September 2008 the UN secretary general in its report observed that "the overall situation in Afghanistan has become more challenging." Despite enhanced capacities, he notes that the " security situation has deteriorated markedly." Further, "the influence of the insurgency has expanded beyond traditionally volatile areas" and "incidents stemming from cross border activities from Pakistan have increased significantly in terms of numbers and sophistication." In addition, he mentions that the humanitarian situation has also deteriorated. The report also draws particular attention to the increase in the number of civilian casualties that "are caused mainly by anti-government activities but are also the unintended consequence of operations by pro-government forces."[17]

In 2009, General Stanley A. McChrystal, the then commander of International Security Assistance Force (ISAF) and U.S. forces in Afghanistan, had warned that the situation is serious and "neither success nor failure can be taken for granted" and that "many indicators suggest that situation is deteriorating." He further wrote that "we face not only a resilient and growing insurgency, there is also a crisis of confidence among Afghans—in both their government and in the international community—that undermines our credibility and emboldens the insurgents."[18]

Reconstruction

Overall, more than 70 nations have committed over US$57 billion for Afghan reconstruction. As of February 2009, the United States had pledged US$38.6 billion, out of which US$22 billion had already been disbursed. As table 12.3 shows, more than 50 percent of this aid has gone into building the Afghan National Army and the Afghan National Police. Other commitments are in the areas of economic and social development, governance, counter-narcotics and support to many civil society activities. Apart from this, the number of U.S. troops serving in Afghanistan could touch 100,000 in 2010.[19] Figures provided in table 12.3 do not include resources provided for military operation in the country.

The other major commitment to Afghanistan is from Europe. Individual member states of the EU and the European Commission are making significant contribution to security and justice reforms, development and reconstruction, counter-narcotics and regional cooperation activities in Afghanistan.

Table 12.3 U.S. Government Funding Provided in Support of Afghan Security, Stabilization, and Development, Fiscal Years 2002–2009

Dollars in millions	*Fiscal Years*								
	2002	*2003*	*2004*	*2005*	*2006*	*2007*	*2008*	*2009ᵃ*	*Total*
Security	**$147**	**$388**	**$949**	**$2,307**	**$1,989**	**$7,431**	**$2,763**	**$5,606**	**$21,580**
—Afghan National Army	86	361	719	1,633	736	4,872	1,778	4,043	14,228
—Afghan National Police	24	0	160	624	1,217	2,523	964	1,512	7,024
—Other security	37	27	70	50	36	36	21	51	328
Governance, rule of law, human rights	**110**	**97**	**262**	**244**	**110**	**286**	**517**	**824**	**2,450**
—Democracy/ Governance	103	89	233	223	80	221	391	614	1,954
—Rule of law	7	8	29	21	30	65	126	210	496
Economic and social development	**650**	**498**	**1,153**	**1,570**	**1,007**	**1,591**	**2,100**	**2,448**	**11,017**
—Reconstruction	124	295	855	1,240	706	1,191	1,494	1,871	7,776
—Humanitarian/ Other	526	203	298	330	301	400	606	577	3,241
Counternarcotics	**40**	**3**	**126**	**775**	**420**	**737**	**617**	**802**	**3,520**
—Eradication	39	0	50	257	138	177	183	202	1,046
—Interdiction	1	3	76	338	137	323	248	366	1,492
—Alternative development	0	0	0	175	140	229	181	225	950
—Other counternarcotics	0	0	0	5	5	8	5	9	32
Total	**$947**	**$986**	**$2,490**	**$4,896**	**$3,526**	**$10,045**	**$5,997**	**$9,680**	**$38,567**

Source: Afghanistan: Key Issues for Congressional Oversight, Report to Congressional Committee (Washington DC: U.S. Government Accountability Office, April 2009), p. 4. Available at www.gao.gov/cgi-bin/getrpt?GAO-09-473SP (accessed November 15, 2009).

EU has also deployed a police mission. Together they have committed around 8 billion euros (around US$11.5 billion) for reconstruction activities. In addition, twenty five EU nations are participating in the NATO-led ISAF with around 30,000 troops. Their military involvement in Afghanistan has been controversial at times because of limits of their deployment and "national caveats" on many of their troops. Many EU nations committed themselves thinking that it would mainly be a peacekeeping and reconstruction effort rather than a project of "war on terror" in which they have to engage with the resurgent Taliban. There have also been problems of coordination with other international partners as well as within the EU nations themselves.

Despite significant success in many areas (economy, education, infrastructure, health, women rights), defeatism is spreading in many European capitals. Exit strategies are being worked out, including "negotiated settlements" with the Taliban. On reconstruction, the United Kingdom has spent over BP 740 million (around US$1175 million) in the last eight years and committed to more than BP 510 million (US$810 million) over the next four years.[20] Germany has also increased its funding in 2008–2009; by 2010, it is likely to have made available resources worth some 1.2 billion euros (US$1.72 billion).[21]From Asia, Japan has pledged around US$1.8 billion to Afghanistan for projects in areas including reconstruction (US$919 million), security (US$212 million), and governance (US$247 million). Together with the UN, Japan is a lead nation in disarmament, demobilization, and reintegration (DDR) and is also involved in the construction of the Kabul-Kandahar highway and a terminal at the Kabul International Airport.

Indian Role in Reconstruction

With a broad understanding that peaceful and stable Afghanistan is crucial for regional stability, India has been playing an active role in the reconstruction since 2002. So far it has pledged assistance for about US$1.3 billion, with projects covering the whole country mainly in the areas of road construction, power transmission lines, hydroelectricity, agriculture, telecommunication, education, health, and capacity building. Details of these projects can be classified under four major heads:[22]

One of the major infrastructural projects completed by India is the construction of the 218-kilometer-long Zaranj-Delaram road in southwestern Afghanistan. This road has a strategic significance for India as it is going to facilitate movement of goods and services from Afghanistan to the Iranian border and, onward, to the Chahbahar Port. This road, together with 60 kilometers of inner-city roads in Zaranj and Gururi, was completed in January 2009 at a cost of US$150 million. During construction 6 Indians and 179 Afghans lost their lives due to insurgent attacks. Another major project, which was completed in 2009, was the construction of a 220kV DC transmission line from Pul-e-Khumri to Kabul and a 220/110/20 kV substation at Chimtala. Built at the cost of US$120 million, this line has facilitated an almost 24-hour power supply from the northern grid to Kabul City. Further, Indian engineers will also be setting up additional 220/20 kV substations at Charikar and Doshi along with Pule-e-Khumri Kabul transmission line. With India's help, construction and commissioning of the 42 MW Salma Dam power project on the Hari Rud River in Herat province is also going to be completed by the end of 2010 at the cost of US$184 million.

The Indian government is also going to invest US$180 million to construct the Afghan Parliament building by 2011. It has also restored telecommunication infrastructure in 11 provinces and expanded national TV network by providing an uplink from Kabul and downlinks in all 34 provincial capitals. Earlier, it also supplied vehicles (400 buses and 200 minibuses for mass urban transportation, 105 utility vehicles for municipalities) and 3 airbus aircrafts and spares to Ariana Afghan Airlines. The Indian government also supplied equipment for three substations in the Faryab province and for a 125-kilometer-long transmission line from Andhkhoi to Maimana, besides rehabilitating Amir Ghazi and the Quargah Reservoir Dam. It further helped in the restoration/revamping of the Afghan media, including the setting up of Azadi (Freedom) printing press, a 100KW–SW transmitter at Yakatoot (Kabul), as well as a TV satellite uplinking/downlinking facility for 10 TV stations and a downlinking facility and TV transmitters in 24 provinces. Other infrastructure projects include solar electrification of 100 villages, construction of a 5000MT cold storage in Kandahar, establishment of a modern TV studio and a 1000W TV transmitter in Jalalabad, setting up of a mobile TV satellite uplink and five TV relay centers in Nangarhar, digging 26 tube wells in 6 northwestern provinces, drilling of 24 deep wells in Herat, planning the construction of a Radio Television Afghanistan (RTA) building in Jalalabad and leasing of slots on the Indian satellite INSAT3A for RTA telecast since 2004.

At the humanitarian level, the Indian government is providing a daily supply of 100 grams of fortified, high-protein biscuits to nearly 1.2 million children under a School Feeding Program. This program is administered through the World Food Program and will cost US$460 million when completed in 2012. It has also reconstructed the Indira Gandhi Institute of Child Health in Kabul and provides free medical consultation and medicines through branches of the Indian Medical Mission in Kabul, Kandahar, Jalalabad, Herat, and Mazar-e-Sharif to over 300,000 patients annually. Apart from supplying blankets, tents, medicines, vegetables, seeds, and other items during 2002–2004, the Indian government also announced a gift of 250,000 metric tons of wheat in 2009.

India is also playing an important role in the field of education by providing 675 long-term university scholarships annually. These fellowships are sponsored by the Indian Council for Cultural Relations for undergraduate and postgraduate studies. In addition, 675 annual slots for short-term technical training courses are being provided every year since 2006. At the January 2010 London Conference, the Indian external affairs minister announced further 200 graduate and 100 postgraduate/PhD fellowships for 5 years in agriculture and related fields. In 2005, with India's assistance, the Habibia School in Kabul was reconstructed and about 9,000 educational kits provided to its students. Further, it provided 20,000 school desks to the Ministry of

Education and laboratory equipments and sports goods to schools in Nimroz as well as teacher training and books to Kandahar and Khost Universities.

In cooperation with the UNDP, the Indian government has also been deputing 30 Indian civil servants as coaches and mentors annually under the Capacity for Afghan Public Administration program since 2007. It has also provided services of Indian banking experts to Da Afghan Bank and the Millie Bank as well as of Indian English teachers in 5 cities; it has provided vocational training in carpentry, tailoring, welding, masonry, and plumbing (through the Confederation of Indian Industries) to 1,000 Afghans as well as in garment making, nursery plantation, food processing, and marketing to 1,000 Afghan women (through the Women's Vocational Training Centre in Baghe-Zanana), besides setting up computer training centers as well as Hindi and English departments at the Nangarhar University. Special training courses have also been provided to more than 150 Afghan diplomats at the Indian Foreign Service Institute, 30 staff of the National Assembly at the Bureau of Parliamentary Studies and Training, about 300 Afghan police, 60 teachers, 60 doctors and paramedics, 60 Ariana Airlines officials, and 40 officials from the Ministry of Mining and Industry. In addition, Indian institutions are also providing training to Afghans in various fields through training programs organized by many international agencies independently.

With the help of the Indian government, around 100 small development projects in the areas of agriculture, public health, rural development, and education have also been under different stages of implementation in 19 provinces of Afghanistan since 2007. In 2002, India contributed US$10 million to the Afghan government budget and has supported the Afghan Reconstruction Trust Fund regularly since 2002. In 2005–2006, it also provided 150 trucks, 15 ambulances, 120 jeeps, bulletproof jackets, bulletproof helmets, laser-aim points, mine detectors, winter clothing, medicines, and other items to the Afghan National Army. It also helped in the setting up of the Common Facilities Service Centre and Tool Room at the Pule-e-Charkhi Industrial Park and trained 5,000 self-help groups in Balakh. The government of India has also agreed to restore the Stor Palace in the Ministry of Foreign Affairs and the House of Screens in old Kabul City.

It seems the Afghans have a very positive perception of Indian activities in the country. This fact has been brought out by many opinion polls. In the latest nationwide survey conducted by the BBC, ABC News, and the German news agency ARD in December 2009, 71 percent of Afghans had a very favorable or favorable opinion about India (see tables 12.4 and 12.5). Corresponding figures for the United States, the United Kingdom, Iran, and Germany were 51 percent, 39 percent, 40 percent, and 59 percent respectively. Only 15 percent of the Afghan population had a favorable opinion

Table 12.4   Afghan Population's Opinion about Different Countries, 2009

	USA	UK	India	Iran	Germany	Pakistan
Very favorable	8	7	29	18	17	2
Somewhat favorable	43	32	42	32	42	13
Somewhat unfavorable	21	28	22	25	21	32
Very unfavorable	25	24	14	20	14	49
No opinion	3	9	7	6	8	3

Source: BBC, ABC and ARD Poll, December 2009. Available at www.news.bbc. co.uk/2/shared/bsp/hi/pdfs/11_01_10_afghanpoll.pdf (accessed January 15, 2010)

Table 12.5   Afghan Population's Opinion about Overall Role Played by Different Countries, 2009

	USA	UK	India	Iran	Russia	Pakistan
Positive	45	28	36	23	22	9
Neutral	18	31	44	29	38	13
Negative	31	31	13	39	31	73
No Opinion	6	10	6	5	5	5

Source: BBC, ABC and ARD Poll, December 2009. Available at www.news. bbc.co.uk/2/shared/bsp/hi/pdfs/11_01_10_afghanpoll.pdf (accessed January 15, 2010)

about Pakistan. Similarly, 44 percent Afghans think that India is playing a neutral role and 36 percent think it is playing a positive role.

Chinese Involvement in Reconstruction and Development

China condemned Soviet intervention in Afghanistan[23] and later cooperated with the United States in arming Afghan Mujaheedin groups against Soviet occupation.[24] After the collapse of the Taliban, Chinese authorities showed relatively little interest in Afghanistan's reconstruction. According to Chinese government sources, China has provided more than 900 million Yuan (US$132 million) in grants to Afghanistan. The main projects include the Jomhuri Hospital and the Parwan Irrigation Project as well as training for about 500 afghan officials in diplomacy, trade, finance, agriculture, counternarcotics, and other fields.[25] It remained disengaged in the country until the Afghan administration opened its energy, mineral, and raw material to foreign investors.[26] In November 2007, the Metallurgical Construction Corporation of China (MCC) and the Jiangxi Copper Limited (JCC) consortium were selected as preferred bidders for the Aynak copper mine project by the Afghanistan Ministry of Mines. In 2008, the Afghan cabinet approved the project.

The Aynak copper mine, a 28 sq. km field, is in Logar province, some 60 kilometers southeast of Kabul. This is the second largest copper mine in the world. According to the Afghanistan Investment Support Agency, MCC plans to invest US$2.9 billion in the project; with investment reaching to US$5 billion in the future. MCC will pay to the Afghan government US$400 million annually to operate the mine and an additional US$800 million as assurance to start extraction work at the copper mine. MCC has also agreed to build a 400 MW power station that will be used both for the mine and for the residents of Kabul. MCC will excavate the coal mine deposits in the area for its energy resource and is committed to build a railway line from the Logar province to one of Afghanistan's borders to facilitate the export of copper. It is expected that this large investment including subprojects will create direct and indirect employment opportunities for 15,000 people. Based on facts from the Ministry of Mines, the Aynak copper mine is believed to contain almost 20 million tons of copper. The construction period of this project is 5 years with a production period of 30 years. There have been allegations that a bribe of US$30 million was paid to get this contract.[27] However, the minister of mines rejects all bribery allegations, saying that MCC's bid was the strongest.[28] In March 2009, at the special conference on Afghanistan, organized under the auspices of the Shanghai Cooperation Organization, members "stressed the importance of concerted and combined efforts in the region to counter terrorism, illicit narcotics and organized crime, in particular coordination and cooperation of Afghanistan and its neighbors."[29]

In recent years, the Chinese have begun to understand that, as a major power, China has to take some major responsibilities in Afghanistan. Broadly speaking, three views of Chinese involvement have emerged. The first view is that China should stop seeing the Afghan issue as an exclusive American problem as it has long-term security implications for China. A second more cautious view indicates many negative consequences for its involvement in the Afghan problem. A third view suggests that instead of a regular army China may send police and paramilitary forces into Afghanistan.[30] Whatever position the Chinese government takes in future, it is becoming clear that it will be more involved in Afghanistan than hitherto.

Regional Economic Cooperation

Despite difficult security situation, bad governance, and limited capacities, Afghanistan could emerge as an important player in regional economic cooperation. Policymakers in Afghanistan believe that after decades of war the country now has a unique opportunity to realize its potential as a "land bridge" between Central, South, and West Asia. They also advocate that peace and stability in this strategically important country is going to provide huge economic opportunities not only to Afghanistan but also to its neighbors. Increasingly it is pointed out that with enhanced cooperation, landlocked energy-rich Central Asia could be connected to energy-deficient South Asia. Similarly, Afghanistan could also realize significant revenue as transit fee and improve its economic activities in the process.

Most official declarations indicate that Afghanistan is seriously committed to regional cooperation. It intends to share the benefits of its centrality through regional cooperation with its neighbors and countries beyond its immediate neighborhood. The two major documents *The Afghanistan Compact* and *The Afghanistan National Development Strategy* clearly show that regional economic cooperation is one of the main priorities of the government. Despite difficult conditions and limited capacities, the Afghan government has been able to market itself as an important player in regional cooperation. This is evident through various international declarations such as the Good Neighborly Relations Declaration of 2002, the Dubai Declaration of 2003, the Berlin Agreements of 2003, the Bishkek Conference of 2004, the Kabul Conference declaration of 2005, the New Delhi Conference declaration of 2006, and the Islamabad Conference declaration of 2009. Through these announcements, countries in the region have accepted the centrality of Afghanistan in economic cooperation.

Afghan Engagement with Regional Organizations

Afghanistan has requested for accession to the World Trade Organization (WTO). In December 2004, the General Council of the WTO established a working party to examine its membership. At the WTO General Council, Afghan ambassador Assad Omer reiterated that Afghanistan "hoped to re-establish itself as the land bridge for trans-continental trade." Apart from making efforts to join the WTO, the country is also simultaneously engaged with many of its neighbors through bilateral and multilateral trade, economic, and investment agreements. Afghanistan is an active member of the Economic Cooperation Organization (ECO). At the fourteenth South Asian Association for Regional Cooperation (SAARC) summit, which was held in New Delhi in April 2007, Afghanistan became the eighth member of the group. Afghanistan's membership in the SAARC has the potential to bring new dynamism in economic relations between the South Asian and Central Asian regions. Afghanistan also serves in the contact group of the Shanghai

Cooperation Organization (SCO). Through its membership in the Central Asian Regional Economic Cooperation (CAREC), the Central and South Asia Transport and Trade Forum (CSATTF), and the UN Special Program for the Economies of Central Asia (SPECA), Afghanistan is also trying to involve itself with various regional projects and activities associated with enhanced regional cooperation. Officially it is claimed that "through regional cooperation, Afghanistan wishes to (a) improve trading opportunities; (b) integrate itself with the regional rail and road networks; (c) be an important partner in regional energy markets; (d) eliminate narcotics trade; and (e) achieve Millennium Development Goals."[31]

Regional Economic Cooperation Conference (RECC) on Afghanistan

To publicize the concept of Afghanistan's centrality and to identify some regional projects, Afghanistan initiated an institutional mechanism called Regional Economic Cooperation Conference (RECC) in 2005. The first RECC was held in Kabul on December 4–5, 2005 and was organized at the initiative of the United Kingdom, the G-8 chair at the time. The Kabul Conference attempted to bring together 11 regional countries—namely, China, India, Iran, Kazakhstan, Kyrgyzstan, Pakistan, Tajikistan, Turkey, Turkmenistan, the United Arab Emirates, and Uzbekistan—and G-8 representatives along with officials from the World Bank, Asian Development Bank, IMF, ECO, SCO, and other such organizations. The focus of the initiative was to promote specific forms of economic cooperation in areas of critical concern to the regional countries. These included specific sectors such as trade, investment and transport facilitation, electricity trade, and energy development, among others. At the end, an ambitious Kabul Declaration was adopted that incorporated decisions on areas that were identified as promoting cooperation. These areas included (a) electricity trade and power development; (b) sharing the benefits of water; (c) counter-narcotics; (d) transport; (e) energy transport; (f) trade facilitation; and (e) business climate.

The second RECC was hosted by India in New Delhi in November 2006. This meeting tried to build on the work done at Kabul in 2005 and followed on the themes identified in that conference. The countries that participated at the New Delhi meeting were Canada, China, Finland (EU presidency), France, Germany, Iran, Italy, Japan, Kazakhstan, Kyrgyzstan, Pakistan, Russia, Tajikistan, Turkey, Turkmenistan, the United Arab Emirates, the United Kingdom, the United States, and Uzbekistan. In addition, ADB, Aga Khan Development Network, European Commission, European Council, IMF, ECO, SAARC, UNDP, UN, and World Bank also participated. Some of the major themes discussed in New Delhi were trade and transport facilitation, investment, regional energy trading, and the Turkmenistan-Afghanistan-Pakistan-India (TAPI) gas pipeline. Special focus was also given to renewal energy and agriculture. In association with the Afghanistan Investment Support Agency (AISA), concurrent business meetings were also held. Some of the recommendations of these meetings included the following:

- The creation of a Centre for Regional Cooperation in Kabul.
- Public-Private partnership as a key aspect of regional development.
- Existing regional groupings should consider integrating their efforts into a larger single entity.
- Afghanistan would benefit more from a region-specific approach to capacity development.
- New themes for the next conference were identified as mining, water, health, labor movement and human resource development, and trade and transit.

Similarly, among other things, the New Delhi Declaration decided that

- Regional countries will undertake stronger credibility/confidence-building measures and will intensify efforts to remove obstacles to overland trade and transit between countries and regions.
- The short-term focus of regional economic cooperation will be on practical win-win projects, notably in the fields of energy, transport and trade, agriculture, and mining.
- There will be better information sharing, via prioritization of key issues, among the countries and regional organizations in order to avoid duplication of efforts.
- ADB will undertake a study on regional integration strategies and will share key findings at the next conference.
- Regional countries will work toward aviation liberalization for greater regional connectivity.
- Work will be accelerated on TAPI gas pipeline to develop a technically and commercially viable project.
- The regional countries will encourage forging of institutional linkages with training institutes in their respective countries with a view to capacity building of their workforce, with the long-term objective of establishing regional training institutes in specialized areas.

At the New Delhi conference it was decided that the next RECC will take place in Islamabad in 2007. However, due to political changes and other reasons, it was postponed many times. Finally, the third RECC took place in Islamabad in May 2009 where many issues concerning trade, energy, capacity building, agriculture, and counter-narcotics were discussed. Among many other decisions, it was agreed that priority will be given to the following items.

- Conclusion of Trade and Transit agreement between Afghanistan and Pakistan before the end of 2009.
- A pre-feasibility study (to be conducted by the European Commission) of railways across Afghanistan linking major destinations within Afghanistan and its neighbors.
- Establishment of a Customs Academy in Kabul.
- Feasibility studies for the development of border economic zones around Afghanistan.
- A centre (supported by the European Commission) within the Ministry of Foreign Affairs to follow up on issues of regional economic cooperation.

At the conference it was also agreed that the next RECC will take place in Turkey.

Developments Targeting Regional Integration

It is not only that Afghanistan and its neighbors have signed some agreements; there is also significant development in the actual realization of some of these initiatives.

Afghanistan has practically no rail or water transport connections. Besides using the meager air transport, the country relies mainly on road transport. Till 1980, Afghanistan had 18,000 kilometers of road network, out of which a stretch of only 3,000 kilometers was asphalted. As a result of decades of conflict, the road network was completely destroyed. In the last few years, the country has been working on a major program of improving its road network. As per the Road Master Plan, Afghanistan has four kinds of roads: national highways (3,363 kilometers), regional highways (4,884 kilometers), provincial roads (9,656 kilometers), and rural roads (17,000 kilometers). For regional cooperation, improvement in regional roads is very important as they connect Afghanistan with its neighboring countries, namely, Iran, Pakistan, Tajikistan, Turkmenistan, and Uzbekistan. They also connect neighboring countries with each other. With the help of multilateral agencies and donor countries (mainly USAID, World Bank, ADB, European Commission, Japan, Iran, India, Pakistan, among others), most of these regional roads have been rehabilitated and/or constructed. Under this program rehabilitation of the ring road has been given priority as it connects the country starting in Kabul, going through Doshi-Puekhumry, Mazar-e-Sharif, Faryab, Badghees, Herat, and Kandahar, and finally ending in Kabul.

The total length of the ring road is 2,210 kilometers. By June 2009, the Ministry of Public Works claimed that 90 percent of the ring road project was complete. The remaining 10 percent was expected to be completed in the next 18 months with an additional cost of US$ 300 million. The roads that connect Afghanistan to its neighbors and consequently to countries in the region are as follows: Kabul-Torkham, Herat-Torghundi, Herat-Islamqala, Kandahar-Spinboldok, Aquina-Andhkhoy, Delaram-Zeranj, and Pule Khumri-Sherkhan Bandar Naibabad. Total length of these roads is 1,153 kilometers. Out of which, 597 kilometers have been rehabilitated so far.

As a significant portion of the national ring road has already been completed, transit time through Afghanistan is greatly lowered. The opening of Sher Khan Bandar "Friendship Bridge" connecting Afghanistan and Tajikistan was also another major milestone. This bridge, 672 meters long and 11 meters wide, and costing US$37 million, will not only connect two neighboring countries but also help increase trade and investment flows in the entire region. The plans for connecting Afghanistan with Iran (Herat-Sangan project) and Pakistan (Chaman-Spinboldok railway) are at different stages of implementation. With the completion of the ring road and further connections with neighbors, Afghanistan can also plan to become a partner in north–south transport corridors. Improvement in transit facilities through Turkmenistan, Tajikistan, and Pakistan and development of Regional Opportunity Zones (ROZ) on the Af-Pak border has the potential to further integrate the region.

In the area of power, Afghanistan is already involved in serious cross-border energy trade. By early 2009, imported power represented about 25–30 percent of Afghanistan's imported power supply. Currently, it imports about 100 MW of power from four neighboring countries, namely, Iran, Turkmenistan, Uzbekistan, and Tajikistan. Under the North Eastern Power System (NEPS), Afghanistan expects to import another 300 MW of power from Uzbekistan by 2008–2009. Similarly, through the Central Asia–South Asia project (CASA-1000), it is expected that by 2012 it will import 1,300 MW of additional power from Tajikistan and Kyrgyzstan; out of this, about 300 MW of power will remain in Afghanistan and the remaining 1,000 MW will be exported to Pakistan. Various bilateral electricity trade deals such as the NEPS,[32] together with evolving new multilateral projects such as CASA-1000, could eventually lead to the creation of an integrated Central Asia–South Asia Regional Electricity Market (CASAREM). In January 2009, a project of electricity transfer from Uzbekistan to Kabul covering a distance of 462 kilometers over the Hindu Kush through the Salang Pass was inaugurated. The 202-kilometer-long 220 KV DC transmission line from Pul-e-Khumri to Kabul along with Chimtala substation project is one of the major Indian infrastructure projects in Afghanistan.

Importance for India

Afghanistan's success in regional cooperation initiatives has major implications for India's linkages with the Central Asian region in the long run. Trade through Pakistan and Afghanistan could also alter India's continental trade. By 2015, India's trade with Europe, CIS plus Iran, Afghanistan, and Pakistan would be in the range of US$500 to 600 billion annually. Even if 20 percent of this trade is conducted through road, US$100 to 120 billion of Indian trade would be passing through the Eurasian region.[33] With improvement in India-Pakistan relations, an important portion of Indian trade (particularly from the landlocked northern states including Jammu and Kashmir) will be moving through Pakistan and Afghanistan. With the possibility of this trade passing through Afghanistan and Central Asia, most of the infrastructural projects in the region will become economically viable. These linkages will also transform small and medium industries and agriculture in Central Asia with the help of good Indian, as well as Pakistani, expertise in these fields. For this to happen, first of all a massive effort is needed to rebuild Afghanistan's transport network and economy. From the commitments of international community and multilateral institutions, it seems that this would happen immediately once there is relative political stability in Afghanistan. The second major impediment in realizing this potential is the existing difficult relation between India and Pakistan. While looking at the regional economic dynamics, it is clear that both India and Pakistan would be paying huge economic costs for not cooperating in the Central Asian region. If trade stops in Pakistan, many road and other infrastructural projects will never become viable because of low volumes. Similarly, India may never be able to radically restructure its continental trade through north-south corridor. Direct linkages between Central Asia and India will also give a huge boost to all economies in the region, particularly in Afghanistan. In cooperation with each other both India and Pakistan could become significant players in Central Asia. The economic costs of continuing conflict is going to be much bigger for both India and Pakistan than normally perceived by policymakers on both sides.

Success in Regional Economic Cooperation
May Lead to Security Cooperation

Compared to a regional approach to economic development that has been well appreciated by all concerned players, the situation in security matters is more complex. Almost all forty-two nations that contribute troops to the ISAF are from outside the region. There are no troops from Iran, China, Pakistan, India, Russia, or Central Asian republics. These are the countries that are actually going to be directly affected by developments within Afghanistan.

Pakistan's involvement in security matters in Afghanistan is complicated. On the one hand, it helps the coalition forces in Afghanistan through logistics, intelligence, joint operations, and other such assistance as a major non–NATO ally. On the other hand, most of the Afghan insurgent groups have bases in Pakistan. In addition, they are given tactical support by many elements within the administration. Regional countries are not involved even in the training of security personnel. This despite the fact that trainers from Iran, India, Pakistan, and the Central Asian region could communicate in local languages and will have more cultural understanding of the place.

Today, what Afghanistan is facing is a typical insurgency. It has broadly two distinct insurgencies. One is the Kandahar-based insurgency that is mainly dominated by the Taliban in the south. The eastern insurgency is more complex. It is a loose confederation of affiliates such as the Haqqani Network and like-minded groups (Al Qaeda, Hizb-e-Islami Gulbuddin, and Pakistani militant groups Jaish-e-Mohammed, Lashkar-e-Tayyiba, and Tehrik Nefaz-i-Shariat Muhammad). Both the Taliban and the eastern groups have support structures in Pakistan. Their shared goals include the expulsion of all foreign military forces from Afghanistan, the elimination of external government influence in their respective areas, and the imposition of a religiously conservative, Pashtun-led government.[34]

On the basis of 90 insurgencies over the world since 1945, a recent RAND study[35] shows that it takes an average of 14 years to defeat an insurgency. Due to history and topography, it might take even longer in Afghanistan. Experience in the region also shows that a large number of boots on the ground are needed to mange any insurgency. The ratio in South Asia is somewhere between 1 to 30 or 35. Current estimates of Taliban and other insurgents in Afghanistan are between 10,000 and 17,000. It means that to manage and defeat this insurgency a minimum 300,000 troops are needed in Afghanistan for at least ten to fifteen years. With the current number of troops it is going to be a very difficult task to manage Afghan insurgency. In the absence of the required numbers, the coalition forces have relied more on airpower in counterinsurgency operations. According to a human rights report, 116 civilians were killed in 2006 in 13 bombings. Similarly, 321 civilians were killed in 2007 in 22 bombings while hundreds more were injured. In the first seven months of 2008 at least 119 civilians were killed in 12 airstrikes.[36] According to UNAMA, in the first six months of 2009, 1,013 civilians died in Afghanistan due to the conflict. Out of this, 59 percent (595 deaths) were caused by antigovernment forces and 310 deaths by pro-government forces.[37] The large number of civilian casualties further undermines the legitimacy of the Afghan administration.

Moreover, it is highly unlikely that many of the ISAF-contributing European nations will be able to commit their forces for the next ten to fifteen years. Many European nations, including Germany and Italy, are seriously debating their involvement in Afghanistan. So strategy in Afghanistan should be to raise the strength of the Afghan National Army to around 200,000 and to involve some of the regional countries in the management of the nation's security. To give a concrete shape to this plan, two major things need to happen in the region. First, a rapprochement between the United States and Iran must take place. Second, a dramatic improvement in relations between India and Pakistan is also a prerequisite for this strategy. Therefore, the war in Afghanistan is not only about troop numbers but also about serious diplomatic engagement by the United States to create a regional platform from where this war will be fought both militarily and diplomatically. Writing agenda for the new U.S. president, the current U.S. special envoy to the Af-Pak region Ambassador Richard Holbrooke wrote last year that "Afghanistan's future cannot be secured by a counterinsurgency effort alone; it will also require regional agreements that give Afghanistan's neighbors a stake in the settlement. That includes Iran—as well as China, India, and Russia. But the most important neighbor is, of course, Pakistan, which can destabilize Afghanistan at will—and has. Getting policy toward Islamabad right will be absolutely critical for the next administration—and very difficult"[38] So far there has been no regional institutional mechanism to address this issue. What is needed is a sustained institutional forum within the region to continue with this engagement on a long-term basis. Half-day meetings in some European capitals will not serve any purpose except for some photo opportunities.

Conclusion

Despite major challenges, Afghanistan has the potential to play an important role in facilitating regional integration for the different economies of South and Central Asia as well as the Middle East. Both India and China are playing a very important role in the reconstruction and development activities of Afghanistan. If proposals concerning regional economic cooperation originating from Afghanistan are implemented by other countries in the region, it could ultimately improve chances of peace not only between India and Pakistan but in the entire Eurasian region. In a typical neo-functionalist way, success in regional economic cooperation could ultimately lead to cooperation in security matters. This would also be useful to create any new institutional economic and security structure that may be needed for any post–NATO scenario in Afghanistan.

Notes

1. Jawaharlal Nehru University, New Delhi.
2. G. Bruno, *Rethinking Afghanistan*, Daily Analysis, September 1, 2009, Council on Foreign Relations, available at <http://www.cfr.org/publication/20114/rethinking_afghanistan.html> (accessed October 1, 2009).

3. B. Raman, "Obama's Af-Pak Troika Fails to Deliver," South Asia Analysis Group, paper no. 3433, September 27, 2009, <http://www.southasiaanalysis.org/papers35/paper3433.html> (accessed October 1, 2009).

4. Remarks by the president of the United States on a New Strategy for Afghanistan and Pakistan, March 27, 2009, <http://www.whitehouse.gov/the_press_office/Remarks-by-the-President-on-a-New-Strategy-for-Afghanistan-and-Pakistan/> (accessed April 1, 2009).

5. "Obama Considers Strategic Shift in Afghanistan" *The New York Times,* September 23, 2009, <http://www.nytimes.com/2009/09/23/world/asia/23policy.html> (accessed October 5, 2009).

6. For text of Bonn Agreement see *Agreement on Provisional Arrangements in Afghanistan Pending the Re-establishment of Permanent Government Institutions,* <http://unama.unmissions.org/Portals/UNAMA/Documents/Bonn-agreement.pdf> (accessed July 1, 2009).

7. See UN Security Council resolution number 1383, December 6, 2001, <http://daccess-ods.un.org/TMP/5722597.html> (accessed October 6, 2009).

8. See *Berlin Declaration,* <http://reliefweb.int/rw/rwb.nsf/db900SID/SNAO-634SXS?OpenDocument> (accessed October 12, 2009).

9. See *Afghanistan Compact,* <http://unama.unmissions.org/Portals/UNAMA/Documents/AfghanistanCompact-English.pdf> (accessed February 27, 2010).

10. See *Paris Declaration,* June 12, 2008, <http://unama.unmissions.org/Portals/UNAMA/Documents/Paris-con-Declaration-English.pdf> (accessed September 20, 2009).

11. *Afghanistan National Development Strategy: A Strategy for Security, Governance, Economic Growth, and Poverty Reduction* (Kabul: Government of Islamic Republic of Afghanistan, 2008).

12. *Afghan Update* (Special Health Issue) UNAMA, no. 20, Summer 2009, p. 7.

13. *Afghanistan: 38% Reduction in Child Mortality, 40% Reduction in Maternal Mortality Achieved During 2002–2008,* News Report <http://www.reliefweb.int/rw/rwb.nsf/db900SID/MYAI-7Z555Z> (accessed December 26, 2009).

14. For details see *Afghanistan Opium Survey 2008* (UN Office of Drugs and Crime, 2008); *Afghanistan Opium Survey 2009 Summary Findings* (UN Office of Drugs and Crime, 2009).

15. *The Afghanistan Study Group: Revitalising Our Efforts, Rethinking Our Strategies* (Washington, DC: Centre for the Study of the Presidency, 2008), p. 5.

16. *Breaking Point: Measuring Progress in Afghanistan* (Centre for Strategic and International Studies, 2007).

17. *The Situation in Afghanistan and Its Implications for International Peace and Security,* Report of the UN Secretary General to the General Assembly Security Council, September 23, 2008.

18. S. A. McChrystal, *Commander's Initial Assessment,* U.S. Commander in Afghanistan, August 30, 2009, p. 1, <http://media.washingtonpost.com/wp-srv/politics/documents/Assessment_Redacted_092109.pdf> (accessed March 3, 2010).

19. "Obama to Send 30,000 More Troops," <http://news.bbc.co.uk/2/hi/8388939.stm> (accessed December 10, 2009).

20. See *The UK in Afghanistan,* October 14, 2009, <http://www.number10.gov.uk/Page20961> (accessed December 28, 2009).

21. "Bilateral Relation: Afghanistan," <http://www.auswaertiges-amt.de/diplo/en/Laenderinformationen/01-Laender/Afghanistan.html> (accessed December 26, 2009).

22. Information on Indian projects in Afghanistan is collected from various sources of the Indian Ministry of External Affairs and information gathered from the Indian embassy in Kabul.

23. "China and Afghanistan," <http://www.mfa.gov.cn/eng/wjb/zzjg/yzs/gjlb/2676/t15822.htm> (accessed December 26, 2009).

24. A. Kuhn, "China Becomes a Player in Afghanistan's Future," <http://www.npr.org/templates/story/story.php?storyId=113967842> (accessed December 2, 2009)

25. <http://www.china.org.cn/world/2010–01/27/content_19311944.htm> (accessed January 27, 2010).

26. N. Norling, "The Emerging China-Afghanistan," *CACI Analyst*, May 14, 2008, <http://www.cacianalyst.org/?q=node/4858> (accessed November 29, 2009).

27. "Afghan Minister Accused of Taking Bribe," <http://www.washingtonpost.com/wp-dyn/content/article/2009/11/17/AR2009111704198.html> (accessed December 25, 2009).

28. "Afghanistan Minister of Mines Denies Bribery Allegations," <http://mom.gov.af/uploads/files/English/Afghanistan%20Minister%20of%20Mines%20Denies%20Bribery%20Allegations%20Final%20.pdf> (accessed December 25, 2009).

29. Declaration of the SCO Special Conference on Afghanistan, March 27, 2009, <http://www.fmprc.gov.cn/eng/zxxx/t554808.htm> (accessed December 5, 2009).

30. See C. Raja Mohan, "Great Game: How China Can Help Afghanistan," <http://www.khaleejtimes.com/DisplayArticle08.asp?xfile=data/opinion/2010/January/opinion_January7.xml&section=opinion> (accessed January 9, 2010).

31. For details see regional cooperation section of the Afghanistan Ministry of Foreign Affairs website <http://www.mfa.gov.af/regional-cooperation.asp> (accessed March 3, 2010).

32. NEPS consists of transmission lines of Hairatan (Uzbekistan border)—Pul-e-Khumri to Kabul—Jalalabad and Kabul—Gardeyz (original plan) plus Sherkhanbander (Tajikistan border) to Pul-e-Khumri and Turkmenistan border to Sheberghan—Mazar-e-Sherif—Pul-e-Khumri.

33. For details G. Sachdeva, "India," in S. F. Starr (ed.), *The New Silk Roads: Transport and Trade in Greater Central Asia* (Washington: Central Asia-Caucasus Institute and Silk Road Studies Program, 2007).

34. For details see: *Report on Progress toward Security and Stability in Afghanistan* (Report to the U.S. Congress in accordance with the 2008 National Defense Authorization Act, June 2008).

35. S. G Jones, *Counterinsurgency in Afghanistan* (Santa Monica, CA: Rand Corporation, 2008).

36. *Troops in Contact: Airstrikes and Civilian Deaths in Afghanistan* (New York: Human Rights Watch, 2008).

37. "Afghanistan: Civilian Causalities Keep on Rising, Says UN Report," *Unog*, July 31, 2009, <http://www.unog.ch/unog/website/news_media.nsf/(httpNewsByYear_en)/0A22BB5BFE041B76C125760400343AE3?OpenDocument> (accessed March 2, 2010).

38. R. Holbrooke, "The Next President," *Foreign Affairs*, vol. 87, no. 5 (2008).

PART IV

Revisited Historical Backgrounds, Disputed Religious Modernities

From the Oxus to the Indus: Looking Back at India-Central Asia Connections in the Early Modern Age

LAURENT GAYER[1]

> The land of Hind is a sugar field
> Its parrots all sell sugar
> Its black Hindus are like flies
> In their turbans and long coats
>
> —Mushfiqi[2]

The Great Mughal Jahangir (1569–1627) had a weakness for Central Asian fruits—a weakness that he inherited from his venerable ancestors. Thus, Babur, the founder of the Indian Mughal Dynasty, considered that "better than the Andijan *nashpati,*[3] there is none."[4] On Babur's part, this definitive statement was probably informed by some nostalgic feeling for his lost homeland (Andijan, the largest town of the Ferghana Valley, was Babur's birthplace). Three generations later, the nostalgia might have receded but the passion remained. Although providing fresh comestibles from so distant a region as Central Asia was quite a challenge for the time, Jahangir took pride in receiving melons from Karis as well as grapes and apples from Samarkand.[5] More than the nostalgia of a Mughal emperor for a "homeland" he had never known, this *péché mignon* suggests that by the early seventeenth century, India-Central Asia trade was flowing unhindered and at a relatively high pace. Traders, soldiers, Sufis, poets, and all sorts of adventurers circulated freely between the two regions, which—along with Iran and the Ottoman

Empire—formed "a single domain of circulation, an ecumene with powerful shared cultural values and symbols."[6]

This chapter is an attempt to reassess these economic and societal relations. Drawing its inspiration from recent works of "connected history,"[7] it emphasizes the singularity of the systemic context in which these relations unfolded, which clearly sets them apart from the international relations that are presently being revived between India and Central Asian republics.

This contribution does not aim to settle scores among historians of medieval India and Central Asia, a task well beyond my competence. It is rather a contribution, by a political scientist with a background in international relations, to contemporary debates around the "reengagement" of India in Central Asia.[8] Against the backdrop of media fantasies[9] and diplomatic rhetoric[10] on the current development of a "new Silk Road," I would like to argue here that these invocations of the past are misleading and amount to a form of anachronism based on "fuzzy analogies" between the past and the present.[11] Such misuses of history are common among international relations practitioners and generally feed on counter-analogies: lessons are to be drawn from past "failures" so that they are not repeated. In the case under study here, this form of presentism[12] operates differently and relies upon wishful-analogies: it is the "successes" of the past that should and, supposedly, can be repeated in the present. Despite their diametrically opposite relation to the past, these forms of reasoning are equally ahistorical, relying upon the misconception that history may well repeat itself, for the worse or, here, for the better.

Connected History and the Persianate Ecumene

The primary objective of connected history, a branch of global history focusing on transcontinental relations between empires of the modern age, is to "overcome the national partitioning of historical research, in order to better grasp phenomena transcending inter-state boundaries."[13] By focusing on imperial connections, these historians also aim to "avoid writing history solely from the West."[14] Rather than a proper historiographic "school," connected history amounts to "a practical research method [that] oscillates between detailed studies of local phenomena of *métissage* affecting practical knowledge, material cultures and doctrines of government, and the hypothesis of transoceanic and transcontinental political and religious configurations."[15]

In recent years, this new agenda of research has produced an impressive body of work on the relations between the four great Muslim empires of the early modern world: Ottoman Turkey, Safavid Iran, Shaibanid Uzbekistan, and Mughal India. Focusing on diplomatic missions between these empires,[16] on travel literature,[17] or on migratory flows,[18] these recent works have brought to light the existence of an "inter-imperial system"[19] integrating South, Central, and West Asia from the sixteenth to the eighteenth century. It was not in religion that this inter-imperial system found its unity, but rather in a shared Persianate culture that accommodated Muslims and non-Muslims alike.[20] This high-culture was not only a factor of cohesion at the cognitive level; it also operated at the pragmatic level by facilitating the circulation of various peoples from one unit of the system to the other. Thus, this wide space of circulation was "defined less by political frontiers than by social ones."[21]

It is to the contours and manifestations of this "inter-imperial system" in the political, religious, and economic affairs of Mughal India that I would like to turn now. Although some mentions will be made of the outreach of this system in Central Asia, the view presented here is resolutely Indo-centric. I am aware that this is a serious limitation for a proper understanding of the dynamics at work here, but space constraints as well as my relatively greater familiarity with Mughal India than with Shaibanid Central Asia or Safavid Iran have led me to limit my ambitions and to adopt an Indian standpoint on these imperial connections.

Political Linkages between India and Central Asia under the Mughals

The Mughal Dynasty that ruled over India between the sixteenth and the mid-nineteenth century[22] was established by Zahiruddin Muhammad Babur (1483–1530), a Chaghtai Turk nobleman hailing from the town of Andijan, in the Ferghana Valley of modern-day Uzbekistan. His father, 'Umar Shaykh, was a great-great grandson of Timur, whereas his mother belonged to the lineage of Chingis Khan.[23] At the death of his father, Babur inherited his fiefdom, proclaimed himself *Padshah*, and embarked on a conquest of the old capital of the Timurids, Samarkand. But he was defeated by the Uzbek tribesmen of Muhammad Shaibani Khan (1451–1510) and, in the process, was expelled from Samarkand and later on from Central Asia at large. Chased away from his homeland, Babur went on another conquering spree in the east, but somehow reluctantly as he would have preferred to rule over Turan than over Hindustan.[24] After establishing his supremacy over Kabul, he defeated the Lodhi sultan at the battle of Panipat (1526) and settled down in Delhi, where succeeding Muslim dynasties had been ruling without interruption from the mid-thirteenth century onward.

As a "typical Timurid," Babur was not a tribal leader and could not rely upon traditional networks of mobilization. In order to consolidate his power, he "had to attract followers using charisma that he laboriously acquired in battle."[25] His own achievements were consolidated by his heirs, with such success that "by the seventeenth century Mughul rulers had successfully made that transition that had eluded all their Central Asian forefathers. They had transformed the weak Timurid patrimonial state into a successful Perso-Islamic absolutist version in which they monopolized claims to legitimacy in northern India."[26] In the process, they both drew inspiration and disassociated themselves from their Mongol and Timurid ancestors, displaying with pride their Central Asian roots while professing their resolute Indian-ness.

The Mughals' Central Asian Legacy

Although Mughal power gradually took roots in India, successive Mughal emperors retained strong bonds with Central Asia, a region they referred to as "the cemetery of the great ancestors" (*gur khana-i ajdad-i 'izam*). Most of them displayed "irredentist impulses"[27] and contemplated recapturing Transoxiana, particularly the province of Badakhshan, which was lost to the Uzbeks in 1585.[28] Babur encouraged his son and heir Humayun to recapture Samarkand. According to his biographer Abul Fazl and to his son Jahangir, Akbar was keen to reestablish Mughal sovereignty over Central Asia. Jahangir also claims to have pondered over an invasion of Central Asia shortly after his accession (1605). Shah Jahan, for his part, momentarily controlled Balkh (1646–1647) and toyed with the idea of recapturing Samarkand and Bukhara but had to review his ambitions in the face of Uzbek resistance and logistical constraints. This "failure" of the Mughals in Transoxiana should not be overblown, though. According to some historians, the Mughals were not genuinely interested in annexing Central Asia to their empire: "The oft-proclaimed desire of recovering the Mughal homelands was really used as a diplomatic ploy, for it was never seriously pursued."[29] Thus, after Akbar extended the northwestern frontier of the empire from the Indus to the Hindu Kush, the Mughals seemed to have given up their expansionist agenda in the northwest.[30] Even Shah Jahan's show of strength at Balkh was primarily diplomatic and aimed to teach a lesson to turbulent Uzbek warlords, who had been launching repeated assaults against Kabul in the preceding years.[31] Shah Jahan understood that the value of these impoverished lands was no match to the cost of such military expeditions. Thus, he spent 40 million rupees in the Balkh campaign, "in an attempt to conquer kingdoms whose total annual revenues were no more than several million rupees." Ultimately, "the Mughal search for familial vindication in this region crashed against the harsh realities of distance, scanty resources, and determined local resistance."[32]

If the Mughals' ambition to reconquer their ancestral lands remains open to question, they "were aware of and reflected on their Central Asian heritage."[33] Central Asian influences were particularly manifest in Mughal political culture. Drawing inspiration and legitimacy from their Turco-Mongol ancestry, the Mughals founded a political system that owed more to the laws of Chingis (the *Tura* or *Yassa*[34]) than to the *sharia*, at least in its restrictive legalistic sense. To the Mughals, the *Tura* was a critical and valuable part of their heritage, despite its pagan origins. Thus, Babur suggests in his autobiography that

> My forefathers had always sacredly observed the rules of Chingez. In their parties, their courts, their festivals and their entertainments, in their sitting down and in their rising up, they never acted contrary to the Tura-I Chengizi. The Tura-I Chengizi certainly possessed no divine authority, so that any one should be obliged to conform them; every man who has a good rule of conduct ought to observe it.[35]

According to Indian historian Satish Chandra, "This willingness to treat the *yassa* as a supplement to the *sharia*, and to issue royal edicts (*yarligh*) to modify the *sharia* whenever it suited them, gave a broader, more liberal character to the Timurid state than the states which had preceded."[36] This is not to say that the Mughals were not concerned in any way with the implementation of the *sharia*. They were adamant to protect and implement the "divine law" (*namus-i Ilahi*), a concept that became synonymous with the *sharia* but was understood in a much larger way than its reductive legalistic sense. The primary concern of the Mughals in implementing this "divine law" was "to ensure a balance of conflicting interests, of harmony between groups and communities, of non-interference in their personal beliefs."[37] Thus, despite criticism from a section of the ulema, to the notable exception of Aurangzeb,[38] the Mughals never embarked on a state-sponsored program of conversion to Islam, renounced to collect the *jizya*[39] from non-Muslims (at least between 1564 and 1679), and even went as far as allowing Hindus to destroy mosques to build their own places of worship.[40]

The Turco-Mongol legacy of the Mughals was also exemplified by the mobility of the emperor and his court, particularly in the second part of Akbar's rule, after he deserted Fatehpur Sikri to embark on a nomadic camp-life. This city of tents could shelter up to 100,000 people and was modeled on Akbar's former capital. By its sheer size and magnificence, it was a powerful testimony of Mughal grandeur. And by delinking imperial authority from the previous seats of Muslim power in North India, such as Delhi, it emphasized that "the emperor himself, rather than a physical site, was the capital of the empire."[41]

Mughal emperors also inherited from the Mongols the tradition of absolute loyalty and obedience to the "great chief" (*qa-an*), which was supposed to deter pretenders to the throne from conspiring against the ruler. Conferring absolute authority to the emperor, this tradition also confined the nobility to a position of "servant" (*nokar*).[42] However, the "Turko-Mongol theory of kingship"[43] was not merely imported into Mughal India; it was reinterpreted creatively by successive rulers, starting with Babur himself. The founder of the dynasty was uncomfortable with the idea of shared sovereignty and set a process of centralization of authority around the emperor's persona—thus wielding an authority that could be transmitted only hereditarily. This process was systematized under Akbar, who adopted a twofold strategy to reinforce his authority by formulating an illuminationist theory (*farr-i-izadi*) inspired by Persian Neoplatonic philosophy, which claimed that the Mughal emperor was of divine nature (the legacy of Chingis here providing some arguments to the ideologues and eulogists of Mughal absolute authority[44]), and by developing a royal cult (*tauhid-i-Ilahi*) that was universal enough to appeal to both Hindus and Muslims[45] and established a relation of spiritual dependence between the emperor and his disciples within the nobility.[46] One should avoid teleological reasoning here, so as to emphasize that these evolutions did not follow a linear, unequivocal trajectory. They emerged only "after a period of considerable contestation," which pitted the strongly Indianized Mughals of the Ganga-Yamuna *doab* against their relatives-turned-rivals settled in Afghanistan, who retained a more Central Asian outlook and were more prone to defend Sunni orthodoxy.[47]

Taming the Turani *Nobility*

Under the Mughals, the nobility and the army comprised large numbers of Central Asians of Chaghtai Turkish or Uzbek lineages, who were known at the time as *Turani*. Renowned for their martial expertise, Central Asians were a favorite for the positions of guards and watchmen (*pasbani*), and Turani soldiers were paid twice the amount offered to Indian recruits.[48] Hundreds of Central Asian nobles accompanied Babur during his conquest of India and the Turanis composed the bulk of the nobility under the first Great Mughal. The share of Turanis in Mughal nobility started declining under Babur's son and heir, Humayun, to the benefit of Persian nobles (*Irani*). After his exile in Safavid Iran (1544–1554), Humayun returned to India with 57 nobles, out of whom 27 were Turanis and 21 Iranis.[49] Turani representation in Mughal nobility declined further during the reign of Jalaluddin Muhammad Akbar (1556–1605), after the Turanis were found involved in several revolts against the emperor.

This Central Asian nobility was initially reluctant to settle down in India. After Babur's conquest of Punjab, many of his *beg* (noblemen) were eager to return home. Their discovery of India was far from idyllic and the bulk of them considered India an unrefined land: "Neither (baked) bread, nor the *hamams*, or social intercourse of the kind they were accustomed to were available in India."[50] Babur also found India "unpleasant and disorderly,"[51] but he was determined to make it his home nonetheless. He made clear to his party that he had no plan to return to Kabul and offered to let the most nostalgic of his *beg* to return home, an option that was taken up by only one of them, Khwaja Kalan. Loyalty to Babur was not the only factor that convinced these reluctant settlers to stay in India: Babur had also granted them prosperous fiefs, many of which remained subdued but promised important revenues. With the passage of time, some Uzbek clans acquired huge landholdings and came to challenge the efforts at centralization of the emperor, out of fear that these would translate into a weakening of their position within the empire.

In the early years of his reign, Akbar found support among the Turani nobility against the regent Bairam Khan, a Persian Shia who was perceived as a threat by the orthodox Sunnis from Central Asia, particularly after he appointed a fellow Shia as *sadr* (religious minister). Akbar's Turani foster brother, Adham Khan, played a key role in this conspiracy against Bairam Khan, which paved the way to Akbar's self-aggrandizement. Uzbek commanders such as Zaman Khan and Bahadur Khan also helped Akbar to consolidate his authority and expand the empire, winning decisive battles against Afghan armies contesting to the Mughals the supremacy over north India.

Akbar's relations with the Turani nobility deteriorated in the second half of the 1560s. These Central Asian nobles traced their lineage back to Shaiban, the Uzbek ruler who had expelled Babur away from Central Asia, and they were less prone to unfettered loyalty than other sections of the nobility.[52] Moreover, these Uzbek nobles resented Akbar's "imperious political style,"[53] both in the country at large and within the court. In 1565–1567, Akbar faced a massive revolt orchestrated by these disgruntled Uzbek nobles, who made an attempt to overthrow the emperor and replace him by his half-brother Muhammad Hakim. The revolt was ultimately crushed by Akbar but Mirza Muhammad Hakim, who retreated to Kabul and retained his grip over the Central Asian territories of the empire, remained a threat to Akbar for another decade. Following this revolt, Akbar proceeded to reduce the proportion of Uzbeks within the nobility. In order to do so, he welcomed Irani nobles who were less reluctant to condone the consolidation of the emperor's authority, a reluctance that found an echo in the imperial Persian tradition of the *padshah*.[54] Akbar also granted important *jagir*[55] to Indian Muslims and Hindu princes (Rajputs, in particular). Between 1575 and 1595, among the 184 nobles of highest rank (*mansab*), the Turanis numbered 64 (34.78 percent), the Iranis 47 (25.54 percent), the Hindustanis (i.e., Indian Muslims) 34 (18.48 percent), and the Hindus 30 (16.30 percent).[56] By 1580, the number of Turanis and Iranians within the imperial elite had become equal.[57] In the process of enlargement of the imperial service, the Central Asian nobility was, therefore, reduced to one component among others of a "composite ruling class."[58]

The Indianization of the Mughal state apparatus under Akbar paralleled a reform of imperial institutions that gradually distinguished the Mughal polity from its Central Asian models. Within Timurid polities of the past, military and financial affairs were concentrated in the hands of a single, all-powerful *wazir*. Babur and Humayun preserved this tradition but Akbar embarked on a reform of imperial structures, thus establishing a clear distinction between religious, financial, and military affairs (the respective domains of the *sadr*, the *diwan,* and the *bakshi*), both at the central and at the provincial level (in 1580, the empire was divided into twelve *subah*).

Integrated Societies

As suggested by historian Muzaffar Alam, "The coming of the Mughals to India in the sixteenth century deepened the pre-existing links between India and Central Asia. [...] Material life in both regions was deeply affected by the accelerated movement of goods and people, while institutions of learning, religion and politics in each area bore the imprint of the other."[59] Although this thesis remains contested by other historians such as Mansura Haidar who have suggested that India-Central Asia commercial relations declined from the early sixteenth century onward, a consensus is emerging among historians of the period around the idea that "commercial relations between India and Central Asia continued in this period, and perhaps even at an escalated level."[60]

Overland and Maritime Trade between India and Central Asia

Trade between India and Central Asia followed maritime as well as overland routes. Along with Hurmuz, the Sindhi ports of Thatta and Lahari Bandar were central nodes in the sea routes linking India to Persia and, through Khurasan, to Central Asia. Although European powers tightened their grip over these sea routes during the seventeenth and eighteenth centuries, this maritime trade survived. During the seventeenth century, Armenians became major intermediaries in these commercial operations, due to their trading agreements with the Europeans. Although they did not try to contest European supremacy over these sea routes, successive Mughal emperors patronized this maritime trade and sometimes became directly involved in it. Thus, Shahjahan owned ships in Surat (Gujarat) and *karkhanas* (workshops) in Burhanpur. His son Darah Shukoh and his daughter Jahan Ara also had personal stakes in this maritime trade. Another of Shah Jahan's sons, Aurangzeb (who would later on succeed his father), made an attempt to build a new port in Sind. One of his grandsons, Prince Azim-al-Shan, was for his part accused of monopolistic control over business in the major port of Bengal, Chittagong.[61]

The overland trade between Central Asia and India followed more perilous and bumpy roads. The most common routes went through the Khyber and Bolan passes, continuing westward to Kabul, Kandahar, or Herat, northward to Balkh or Kashgar and from there to Samarkand and Bukhara, eastward to Multan, Lahore, and Kashmir or southward to Baluchistan and Sindh. Although the principal land routes remained unchanged over the centuries, despite their perils and their appalling conditions, political developments dictated marginal changes in their course, these adjustments being enough to precipitate a change in fortune or the demise of entire towns (if not of entire regions) over a brief period of time. Thus, the uprising of the Sikhs in the Punjab in the eighteenth century was to a large extent the outcome of an economic crisis precipitated by the decline of the trade with Central and West Asia.[62]

The major commodities flowing from India to Central Asia included spices, textiles, and slaves (both Hindus and Muslims), as well as indigo, precious stones, sugar, medicinal herbs, and other items. In the other direction, Central Asia provided military horses, dry and fresh fruits, silks, furs, musk, cotton, precious metals, falcons, corals, and other items. The balance of this trade was undoubtedly in India's favor and its scale can be gauged by the number of horses that were imported annually into India from Central Asia.[63] According to some reports, this number could have reached 100,000 in the seventeenth century and remained around 50,000 late into the eighteenth century.[64] Another indication of the sheer size of this overland trade can be found in the outcome of an accidental fire that occurred in the Peshawar Fort in 1586; the fire led to the destruction of 1,000 camel loads belonging to the merchants who had taken shelter there.[65]

Peripatetic Poets and Sufis

During the sixteenth and seventeenth century, Safavid Iran and Shaibanid Central Asia witnessed a "brain-drain" toward Mughal India.[66] One of the largest contingents of this intellectual emigration constituted of poets, who were attracted to the Mughal court by the largesse of successive emperors and their patronage of Persian-speaking men of letters.[67] The founder of the Mughal Dynasty, Babur, was himself an accomplished poet and "traded verses with pen pals such as Alisher Nawa'i of Herat and with Bina'i".[68] These literary contacts between Indian and Central Asian or Iranian poets gave birth to a specific genre of poetry known as *mujavaba*, "in which a person would write a line or two and send it to a friend, who would then write a response in the same rhyme and metre. The mailbags of caravans connecting the

Uzbek and Mughal lands were filled with this type of back-and-forth one-upmanship."[69]

One of these Central Asian poets, Mutribi al-Asamm al-Samarqandi, has left us a detailed account of his visit to Jahangir's court in the 1620s.[70] After singing praises of the emperor in a poem of his composition, Mutribi was presented with gifts (money, a robe of honor, a horse, and a saddle). The conversations between Mutribi and the emperor really started off during their second meeting. Jahangir was eager to question his visitor about the state of affairs in Central Asia. His interests revealed the nature of his bond with Turan. Most of all, Jahangir was preoccupied with the state of the Gur-i-Amir, Timur's mausoleum in Samarkand. By enquiring about details such as the color of the gravestone, Jahangir reaffirmed the attachment of the Mughals to their Timurid ancestry, a fact corroborated by the contribution of successive Mughal emperors to the *waqf* of the Gur-i Amir. In short, Mutribi "represents a window into Central Asia for Jahangir, as a sort of authentic witness (*bayan*) to affairs in Transoxiana."[71] However, throughout his encounters with Mutribi, Jahangir also appears eager to impress his visitor with local wonders unknown in Transoxiana (such as a massive piece of sugar-candy, a camel tournament). Thus, "a persistent thread that runs through the conversations concerns Jahangir's effort to demonstrate the hierarchical superiority of Hindustan over Central Asia."[72] These efforts displayed by Mughal emperors to impress their Central Asian visitors "might to some extent have stemmed from insecurity about losing their original homeland to the Uzbeks and the desire to convey to Central Asians the fact of their 'having made good' in exile."[73]

Mutribi, for his part, was eager to build bridges between the Mughal court and the rulers of Turan. Thus, he reports that he aimed to carry back to Imam Quli Khan some of the gifts that Jahangir had presented him with—in particular a "magical" pencil offered to the emperor by some European merchants.[74] In a time of diplomatic ambiguity, when self-professed ambassadors could be deemed impostors,[75] travelers such as Mutribi were important agents of connection between distant empires with limited official channels of communication. They could entrust themselves with diplomatic tasks (probably expecting to benefit personally from these) or be endowed with official missions. Thus, Mutribi was asked by Jahangir the amount of money that was necessary to repair the Gur-i Amir. The poet suggested that 10,000 rupees would be required for such an endeavor and the emperor committed to sending this amount through Mutribi.[76]

Texts such as Mutribi's were instrumental in instilling new migrants' vocations among sections of the Central Asian nobility or religious circles. Despite his attraction for Indian curiosities and wonders (*'aja'ib*), Mutribi presents his readers with a particularly bland picture of India. What is striking in his account is his utter neglect of the realities of India beyond the Mughal Court. One finds no mention of city or village life in his text. Thus, "the vision presented of Mughal India to the reader in Central Asia is one that [...] remains one of courtly life and hence of a domain where the largely familiar mores of Persianate culture hold sway."[77]

Sufi orders also contributed to the regional integration of Central Asia and the Mughal Empire. The Mughals traditionally patronized the "orthodox", Sharia-oriented Naqshbandis. This politics of patronage replicated that of the Mongols and of earlier Timurids, who had used their proximity to the Sufis as a source of Islamic legitimacy.[78] The knot between the Mughals and this orthodox Sunni order was tied in Central Asia by Babur's family. His father 'Umar Shaykh and his uncle Sultan Ahmad were disciples of the founder of the Naqshbandi order, Khwaja 'Ubayd Allah Ahrar (1404–90), and their devotion to the Ahrari family was transmitted to Babur and his offspring. This intimate relation between Babur and the Naqshbandis provided the former "an essential spiritual and symbolic link between the former Central Asian homeland and the new Timurid-Mughal dynasty he established in India."[79] It also paved the way for the order's development in India, at the initiative of Khwaja Baqi Bi'llah (1563–1603) and Ahmad Sirhindi (1564–1624). Under the guidance of the latter, known to his followers as the "Renewer of the second millennium" (*mujaddid-i-alf-i-thani*), the Naqshbandiya ceased to be confined to the courtly milieu and became a truly Indian order. This local branch of the Naqshbandiya developed institutional and theological singularities[80] and came to be known as the Mujaddidiya. Under the tutelage of Indian Mujaddidi sheikhs settled in Central Asia and of Central Asian Naqshbandi sheikhs who had studied in India, the Mujaddiya was exported in Transoxiana, starting at the turn of the seventeenth century and gaining intensity in the second half of the eighteenth century.

The privileged relationship between the Mughals and the Naqshbandiya, which was reinforced by intermarriages, was temporarily suspended under the reign of Akbar and, to a lesser extent, that of Jahangir, who favored the less orthodox Chishtiya order. However, the bonds between the Naqshbandiya and the Mughals were revived under the reign of Shah Jahan. The latter used Central Asian Naqshbandis as emissaries and informers. Sufi orders also facilitated the integration of Hindu settlers in Central Asia, since many of them were disciples (*murids*) of Sufi pirs.[81] In the opposite direction, the tombs of Sufi saints attracted large numbers of pilgrims and "sightseers" from Central Asia.[82]

Migrants and Refugees

Migratory flows between Central Asia and India during the period under study were essentially eastward. Although a handful of Indians—generally bankers, merchants, or master weavers—settled down in Central Asia in the sixteenth century, if not earlier, their numbers were limited compared to those of Central Asians travelling in the opposite direction.[83] In the Uzbek Khanate, these Hindu migrants (who were almost exclusively male) initially blended into the local population, living in mixed areas, converting to Islam, and marrying local women. In the seventeenth century, they were granted a large degree of autonomy in the management of their daily affairs, which was left to one of their member, the *aqsaqal*.[84] This policy contributed to the development of a new diasporic consciousness among this population, which became more aware and more assertive of its cultural particularities. However, for some authors, the use of the term "diaspora" is inappropriate in this context. Thus, in his study of Sindhi merchants from Shikarpur settled in Central Asia, Claude Markovits prefers to think in terms of "network" due to the chronic ambulatory practices of these merchants between their homeland and Central Asia.[85] Anita Sengupta, for her part, considers these merchants "sojourners" rather than "settlers," for they retained strong connections with their hometowns.[86]

In the opposite direction, the majority of Central Asians trying their luck in India were economic migrants attracted by the country's relative cultural proximity and its supposedly unlimited riches. Indeed, for the peoples of Central Asia, "India was the land of gold and slaves."[87] These migratory flows primarily concerned the elite circles of the nobility, the literati, and the merchants who could expect to benefit from their genealogical, cultural, or economic capital in India. However, less privileged sections of Central Asian societies also travelled extensively between the two regions and played a key role in the overland trade between them. Such was the case, in particular, of the *powindas*, pastoral nomads of Pakhtun extraction, who shared their time between India (where they settled during the winter) and Transoxiana (where they resided during the summer).[88]

Natural calamities also affected the movements of population between Central Asia and India. Thus, 12,000 Central Asians facing starvation are said to have fled to India in the 1730s.[89] These migratory flows were also invigorated by political circumstances that led to subjects of tyrannical rulers of Transoxiana protesting with their feet and migrating to India, such as during the invasion of Balkh by Abdullah Khan.[90] However, according to historian Jos Gommans, one should refrain from interpreting migratory flows between Central Asia and India as "a flight from chaos, stagnation and suppression", as they were primarily "a sign of economic vitality."[91]

Perilous Roads

Piracy, banditry, and political unrest were a constant threat over these flows of goods and men. This insecurity was particularly notable on both sides of the Khyber Pass. Successive military campaigns to subdue Pakhtun tribes, the Yusufzai in particular, met with mitigated success. They gradually extended Mughal authority to these mountainous regions but could not entirely eliminate the aforementioned threats. Thus, the 1585 attempt by Akbar to tame the Yusufzais ended up with "the greatest disaster to Mughal arms in Akbar's reign."[92] In the following years, Mughal armies built an impressive network of forts to secure the caravan trade.

Overland routes between Central Asia were particularly dangerous for individual travelers. For merchants involved in the India–Central Asia trade, the prospects of amassing wealth were as great as those of losing everything to bands of marauders. If they were not killed by bandits, they could be captured and sold as slaves in the bazaars of Central Asia.[93] Thus, Indian and Central Asian traders generally restrained from travelling in small parties and preferred to join their forces with other travelers, even if this implied repeated losses of time at successive *sarai* along the way. Pastoral nomads such as the *powindas* were less exposed to the threat of banditry as they travelled in large groups under heavy military protection.[94]

Conclusion

The movement of goods, ideas, and men between India and Central Asia as well as the concrete interactions between their peoples are remarkable for their intensity and duration. For many present-day commentators, diplomats or otherwise, these past connections could constitute an asset for the revival of economic and diplomatic cooperation between the now estranged regions. Hence the frequent invocation of the legacy of the Silk Road in present-day relations between India and Central Asia, as if present developments could unfold only on the pattern of the past. However fuzzy these analogies may be, they retain a strong symbolic power. Thus, the first consignment of Afghan apples to India, after the fall of the Taliban regime, was branded the "Silk Road harvest."[95]

Notwithstanding their evocative power, these endorsements of past relations between India and Central Asia for the sake of contemporary developments tend to miss the point. The intense relations between these two regions from the sixteenth to the eighteenth century took place within the framework of an "inter-imperial system" that included the Ottomans, the Mughals, the Safavids, and the Shaibanids. Each of the parties to this system considered itself to be endowed with a level of sovereignty comparable to that of other parties and the integration of the four units was facilitated by the prevalence of a common Persianate code of conduct, which transcended religious affiliations and "provided the basis for communication and diplomatic dealings." In such a space of circulation, it was possible for travelers such as Mutribi al-Asamm al-Samarqandi "to arrive in a strange setting, immediately find his bearings, and function with a great deal of ease"—at least provided he remained confined to the courtly milieu.[96]

This inter-imperial system, which was so instrumental in sustaining deep and wide relations between India and Central Asia from the sixteenth century onward, collapsed in the second half of the nineteenth century. The geopolitical anxieties of the parties to the "Great Game," the Russian revolution and the subsequent Sovietization of Central Asia, the decline of the Ottoman Empire, and the de-Persianization of India, all contributed to this changed state of affairs. More recently, the end of the cold war and the beginning of Central Asia's access to independence have consecrated the triumph of the nation state over empire from the Oxus to the Indus and beyond. In the process, national identities have been sharpened and frontiers made more impermeable. In this new international context, it seems illusory to expect relations between the peoples and rulers of the two regions to regain the scope they had within the "inter-imperial system."

Notes

1. Research Fellow, CNRS-CURAPP, Amiens—CSH, New Delhi.
2. Mushfiqi was a well known Central Asian poet of the sixteenth century. Born in Bukhara, he migrated to India in 1567–1568 after an unsuccessful attempt to achieve notoriety in Samarkand; quoted in Richard Foltz, "Cultural Contacts between Central Asia and Mughal India," in S. C. Levi (ed.), *India and Central Asia. Commerce and Culture, 1500–1800* (Delhi: OUP, 2007), p. 159.
3. A kind of melon or a pear.
4. Zahiruddin Muhammad Babur, *Babur-Nama*, translated by Annette Susannah Beveridge (New Delhi: Low Price Publications, 1989 [1921]), p. 3. The following pages of Babur's autobiography, through which he introduces the reader to his Central Asian homeland, are full of references to local fruits (pomegranates, grapes, apricots, apples, melons, etc.).
5. M. Alam, "Trade, State Policy, and Regional Change Aspects of Mughal-Uzbek Commercial Relations, c. 1550–1750," in S. C. Levi (ed.), *India and Central Asia. Commerce and Culture, 1500–1800* (Delhi: OUP, 2007), p. 70.
6. M. Alam, S. Subrahmanyam, *Indo-Persian Travels in the Age of Discoveries 1400 and 1800* (Cambridge: Cambridge University Press, 2007), p. 229.
7. S. Subrahmanyam, *Explorations in Connected History. From the Tagus to the Ganges* (Oxford: Oxford University Press, 2005).
8. For a current state of these debates, see E. Kavalski, *India and Central Asia. The Mythmaking and International Relations of a Rising Power* (London: I.B.Tauris, 2009); L. Gayer, "Le réengagement de l'Inde en Asie centrale," *Politique étrangère*, no. 3 (2008), pp. 589–600.
9. See, for instance, "The New Silk Road," *Business Week*, November 6, 2008; "The New Silk Road," *The Washington Post*, April 09, 2007; P. Escobar, "Iran, China and the New Silk Road," *Asia Times Online*, July 26, 2009.
10. See, for instance, the communicate of the Indian Embassy in Moscow, "India and Russia Will Come Closer through a New Silk Route," June 30, 2004, <http://www.indianembassy.ru/cms/index.php?option=com_content&task=view&id=209&Itemid=626> (accessed March 3, 2010).

11. R. E. Neustadt, E. R. May, *Thinking in Time. The Uses of History for Decision-Makers* (New York: Free Press, 1986), p. 33.

12. On the various forms of presentism in the theory and practice of international relations, cf. Colin Elman and Miriam Fendius Elman, "Diplomatic History and International Relations Theory: Respecting Difference and Crossing Boundaries," *International Security*, vol. 22, no. 1, Summer 1997, pp. 5–21.

13. C. Douki, P. Minard, "Histoire globale, histoires connectées: Un changement d'échelle historiographique—introduction," *Revue d'histoire moderne et contemporaine*, no. 54–55, 2007/5, p. 10.

14. Ibid., p. 11.

15. R. Bertrand, "Rencontres impériales: L'histoire connectée et les relations euro-asiatiques," *Revue d'histoire moderne et contemporaine*, no. 54–55, 2007/5, p. 69.

16. S. Subrahmanyam, "Par-delà l'incommensurabilité: Pour une histoire connectée des empires aux temps modernes," *Revue d'histoire moderne et contemporaine*, no. 54–55, 2007/5, pp. 34–53.

17. M. Alam, S. Subrahmanyam, *Indo-Persian Travels in the Age of Discoveries 1400 and 1800*.

18. S. F. Dale, *Indian Merchants and Eurasian Trade, 1600–1750* (Cambridge: Cambridge University Press, 1994); S. C. Levi, *The Indian Diaspora in Central Asia and Its Trade, 1550–1900* (Leiden: Brill, 2002); S. Subrahmanyam, "Iranians Abroad: Intra-Asian Elite Migration and Early State Formation," *Journal of Asian Studies*, no. 51, 1992, p. 340–363.

19. M. Alam, S. Subrahmanyam, *Indo-Persian Travels in the Age of Discoveries*, p. 129.

20. On the Persianization of Mughal India, see M. Alam, *The Political Language of Islam in India* (Delhi: Permanent Black, 2004), chap. 4, "Language and Power."

21. R. Foltz, "Cultural Contacts between Central Asia and Mughal India," p. 163.

22. Although the Mughals formally ruled over India until 1857, their empire started declining in the early eighteenth century, after the death of Aurangzeb, the last of the "Great Moguls."

23. Despite the fact that Babur's mother was a descendant of Chingis' son Chaghtai, he was never recognized as a "true Chingis" by Central Asians. But this deficit of legitimacy was not as problematic in India as it would have been in Central Asia ; on this point, see M. E. Subtelny, "Babur's Rival Relations: A Study of Kinship and Conflict in 15th–16th Central Asia," *Der Islam*, vol. 66 (1989), pp. 102–118.

24. S. F. Dale, *The Garden of the Eight Paradises. Babur and the Culture of Empire in Central Asia, Afghanistan and India (1483–1530)* (Leiden: Brill, 2004), p. 358.

25. S. F. Dale, "The Legacy of the Timurids," in S. C. Levi (ed.), *India and Central Asia. Commerce and Culture, 1500–1800* (Delhi: OUP, 2007), p. 180.

26. Ibid., pp. 180–181.

27. Ibid., p. 181.

28. M. Athar Ali, "The Objectives behind the Mughal Expedition into Balkh and Badakhshan, 1646–1647," in *Mughal India. Studies in Polity, Ideas, Society, and Culture* (Delhi: OUP, 2009 [2006]), p. 330.

29. Satish Chandra, *Medieval India. From Sultanat to the Mughals*, vol. 2, *Mughal Empire (1526–1748)* (Delhi: Har-Anand, 2009 [1999]), p. 229.

30. The successive battles for Kandahar between Mughal, Safavid and Uzbek forces, all along the seventeenth century, are not at odds with this strategy: Kandahar was simply an outpost of the Mughal Empire, rather than a base for the territorial expansion of the empire toward the west.

31. On the various motives behind this expedition, see M. Athar Ali, "The Objectives behind the Mughal Expedition into Balkh and Badakhshan, 1646–1647."

32. J. F. Richards, *The New Cambridge History of India. The Mughal Empire* (Cambridge: Cambridge University Press, 2007 [1993]), p. 133.

33. M. Alam, S. Subrahmanyam, "Introduction," in M. Alam, S. Subrahmanyam (eds.), *The Mughal State, 1526–1750* (Delhi: OUP, 2000 [1998]), p. 17.

34. For some historians, the two terms were not synonymous in Mughal India: Whereas the *Yassa* referred to Chingis' rules *stricto sensu*, the *Tura* referred to "Turco-Mongol customary practices and royal traditions" in a broader sense. It is under this extensive form that Mongol customs were perpetuated by the Mughals (at least until Shah Jahan's reign), particularly in the court etiquette; cf. M. Haidar, *Indo-Central Asian Relations. From Early Times to Medieval Period* (Delhi: Manohar, 2004), p. 129.

35. *Babur-Nama*, translated by John Layden and William Erskine, vol. 2, Oxford, 1921, quoted by Muzaffar Alam, *The Language of Political Islam in India* (Delhi: Permanent Black, 2004), p. 81.

36. S. Chandra, *Medieval India,* vol. 2, p. 14.

37. M. Alam, *The Language of Political Islam in India,* p. 78.

38. Aurangzeb encouraged conversion to Islam among the imperial elite by granting promotions or distinctions to recent converts; J. F. Richards, *The Mughal Empire,* p. 177.

39. According to the *sharia*, the *jizya* is a protection tax to be collected by Muslim rulers from their non-Muslim subjects.

40. M. Alam, *The Language of Political Islam in India,* p. 77.

41. J. F. Richards, *The Mughal Empire,* p. 12.

42. S. Chandra, *Medieval India,* vol. 2, p. 15.

43. R. Prasad Tripathi, "The Turko-Mongol Theory of Kingship," in M. Alam, S. Subrahmanyam (eds.), *The Mughal State, 1526–1750* (Delhi: OUP, 2000 [1998]), pp. 115–125.

44. Chingis descended from Queen Alanquwa, who according to the myth retold in the introduction of the *Akbar Nama* (Akbar's monumental "biography" by Abul Fazl), had given birth to triplets after she was fecundated by the Sun ; on this basis, Akbar's leading ideologue and biographer, Abul Fazl, as well as his brother the poet Faizi, propagated the idea that Akbar was "the emanation of God's light," to quote one of Faizi's eulogistic quatrains ; see J. F. Richards, "The Formulation of Imperial Authority under Akbar and Jahangir," in M. Alam, S. Subrahmanyam (eds.), *The Mughal State, 1526–1750* (Delhi: OUP, 2000 [1998]), p. 141.

45. Ritually, this new cult (which was perpetuated by Jahangir in the early part of his reign) was based on the worshipping of the sun. As such, it could relate to Rajput traditions and thus strengthen the bonds between Akbar and his Rajput nobles and allies. In the 1580s, Akbar started worshipping the sun openly by prostrating himself four times a day before a fire; see J. F. Richards, *The Mughal Empire,* p. 47.

46. J. F. Richards, "The Formulation of Imperial Authority under Akbar and Jahangir."

47. M. Alam, S. Subrahmanyam, "Introduction," p. 22.

48. B. B. Kumar, "India and Central Asia: Links and Interactions," in J. N. Roy and B. B. Kumar (eds.), *India and Central Asia. Classical to Contemporary Periods* (Delhi: Astha Bharati, 2007), p. 17.

49. Ibid.

50. S. Chandra, *Medieval India,* vol. 2, p. 32.

51. Quoted in S. F. Dale, *The Garden of the Eight Paradises,* p. 371.

52. J. F. Richards, *The Mughal Empire,* p. 17.

53. Ibid.

54. Ibid., p. 19.

55. The grant of a *jagir* did not translate into propriety rights over a specific land but allowed its beneficiary to collect all taxes due to the state.

56. S. Chandra, *Medieval India,* vol. 2, p. 126.

57. J. F. Richards, *The Mughal Empire,* p. 19.

58. Ibid., p. 124.

59. M. Alam, "Trade, State Policy, and Regional Change Aspects of Mughal-Uzbek Commercial Relations," p. 64.

60. S. C. Levi, "Introduction," in S. C. Levi (ed.), *India and Central Asia. Commerce and Culture, 1500–1800* (Delhi: OUP, 2007), p. 7.

61. M. Alam, "Trade, State Policy, and Regional Change Aspects of Mughal-Uzbek Commercial Relations," p. 76.

62. Ibid., p. 83.

63. On the history of this horse trade, see Jos Gommans, *The Rise of the Indo-Afghan Empire, c. 1710–1780* (Leiden: E. J. Brill, 1995), chap. 3, "Horse Breeding and Trade in India."

64. S. C. Levi, "Introduction," p. 10.

65. J. F. Richards, *The Mughal Empire,* p. 50.

66. R. Foltz, "Cultural Contacts between Central Asia and Mughal India," p. 156.

67. Relying upon anthologies of poets of the seventeenth century, Richard Foltz suggests that at least 10 percent of Central Asian men of letters of the time might have sojourned in Mughal India; ibid., p. 159.

68. Ibid., p. 158.

69. Ibid., p. 159.

70. M. Samarqandi, *Khatirat-i Mutribi Samarqandi* (Persian) (The Conversations of Mutribi Samarqandi), Abdul Ghani Mizoyef (ed.), Karachi, 1977, presented and commented in M. Alam and S. Subrahmanyam, *Indo-Persian Travels in the Age of Discoveries 1400 and 1800,* pp. 120–129.

71. M. Alam, S. Subrahmanyam, *Indo-Persian Travels in the Age of Discoveries 1400 and 1800*, p. 123.

72. Ibid., p. 128.

73. R. Foltz, "Cultural Contacts between Central Asia and Mughal India," p. 155.

74. Ibid., p. 123.

75. In 1608, a Central Asian going by the name of Aqam Haji turned out at Jahangir's court, claiming to be an ambassador of the Ottoman sultan. The Mughal emperor and his court suspected that the letter he had brought was a fake and he was turned away; ibid., p. 120.

76. Ibid., p. 128.

77. Ibid., p. 129.

78. J.-A. Gross, "The Naqshbandiya Connection: From Central Asia to India and Back," in S. C. Levi (ed.), *India and Central Asia. Commerce and Culture, 1500–1800* (Delhi: OUP, 2007), p. 233.

79. Ibid., p. 238.

80. Sirhindi propagated the doctrine of *wahdat-al shuhud* (unity of perception) against that of *wahdat-al wujud* (unity of being). Whereas the latter advocated religious tolerance and paved the way for the assimilation of local, pagan cultural traits, Sirhindi's rival doctrine claimed that this apparent unity of being was an illusion. At a practical level, this was supposed to consolidate the boundaries between Islam and Hinduism and to put an end to syncretic beliefs and rituals that had become characteristic of Indian Sufism ; cf. M. Alam, *The Language of Political Islam in India,* p. 163.

81. C. Markovits, "Indian Merchants in Central Asia: The Debate," in S. C. Levi (ed.), *India-Central Asia Relations*, (Delhi: Oxford University Press, 2007), p. 137.

82. R. Foltz, "Cultural Contacts between Central Asia and Mughal India," p. 157.

83. Historians estimate that, in the seventeenth century, this Indian trade diaspora's numbers were in the tens of thousands; *cf.* C. Markovits, "Indian Merchants in Central Asia," p. 125.

84. M. Alam, "Trade, State Policy, and Regional Change Aspects of Mughal-Uzbek Commercial Relations," p. 79.

85. C. Markovits, "Indian Merchants in Central Asia," pp. 128–129.

86. A. Sen Gupta, "Diasporas along the Silk Road: The Indian Trader Communities in Central Asia," in L. Gosh, R. Chatterjee (eds.), *Indian Diaspora in Asian and Pacific Regions. Culture, People, Interaction* (Jaipur/Delhi: Indian Association for Asia Pacific Studies/Netaji Institute of Asian Studies, 2004), pp. 67–90.

87. J. Gommans, "Mughal India and Central Asia in the Eighteenth Century: An Introduction to a Wider Perspective," in S. C. Levi (ed.), *India-Central Asia Relations* (Delhi: Oxford University Press, 2007), p. 43.
88. Ibid., p. 44.
89. B. B. Kumar, "India and Central Asia," p. 17.
90. Ibid., p. 18.
91. J. Gommans, "Mughal India and Central Asia in the Eighteenth Century," p. 43.
92. S. Chandra, *Medieval India,* vol. 2, p. 51.
93. See, for instance, the tribulations of the Indian merchant 'Ala-al Din Khan, who in the mid-seventeenth century was enslaved on his return to India from Balkh ; *cf.* M. Alam, "Trade, State Policy, and Regional Change Aspects of Mughal-Uzbek Commercial Relations," art. quoted, p. 68.
94. Ibid., p. 44.
95. "Afghan Apples Headed for Indian Markets," *Yahoo India News*, November 24, 2009.
96. M. Alam, S. Subrahmanyam, *Indo-Persian Travels in the Age of Discoveries 1400 and 1800*, p. 129.

CHAPTER 14

Uyghur Islam: Caught between Foreign Influences and Domestic Constraints

RÉMI CASTETS[1]

Trade along the Silk Road and ethnocultural continuities[2] between both sides of the Tianshan Range have favored constant intellectual exchanges between western and eastern Turkestan.[3] Thus, during the last millennium, Sufi leaders and preachers originating from Central Asia played an active role in Islamizing the Turkic populations of present-day Xinjiang. As a result, Uyghur traditional Islam resonates with interpretations of the Koran that have emerged in the religious continuum Xinjiang was forming at that time with western Turkestan, northern India, or, even further, Iran.

The decline of commercial exchanges along the Silk Road, the integration of the region into the Qing Empire in the mid-eighteenth century, and the subsequent repression of Sufi networks openly opposed to Manchu/Chinese sovereignty were not much of a threat to those religious and intellectual exchanges. Rather, in the beginning of the twentieth century, life in non-Muslim countries prompted a part of Islamic Uyghur elites to look closer at the centers of Ummah. The feeling of living far from the main currents of Islamic knowledge, the feeling that non-Muslim rule could pervert the society and Islamic practices, and the deep rooting of heterodox syncretisms criticized by foreign intellectuals,[4] all fed the idea among Uyghur reformists that Islam in the region was under threat of degeneration. This fear and the idea that the quintessence of Islamic knowledge was located in the western territories of Dar al-Islam strengthened elites' attraction to schools of thought coming from those regions. Due to the old networks mentioned above, close linguistic affinities, and complications associated with traveling to the Near and Middle East (i.e., time length and costs), in the beginning of the twentieth century, most of the Uyghurs wishing to deepen their knowledge of Islam were drawn toward the prestigious Bukhara or Islamic teachings in the Tatar world.

Starting in that period, those centers acted as interfaces between this isolated region and the modern world. They have favored the diffusion of the syncretisms that Islamic elites exposed to Western modernity[5] brought to the Arabic and the Turkic world. In other words, in that region, strongly influenced by Central Asian intellectual scene, the diffusion of modern ideologies through mass communication and education, the rise of Chinese modern nation-state, and the acceleration of the circulation of foreign interpretations of Islam through present-day globalization processes have all together fed deep transformations of the way Uyghurs conceive Islam.

This chapter pays particular attention to those fundamental evolutions. More specifically, it will try to address how the intrusion of modernity and the restructuring of the political and ideological context in which Uyghurs have lived have modified the status of Islam, the architecture of Islamic currents, and religious practices in twentieth-century Xinjiang.

Religious Spaces and Currents amongst Sedentary Turkic Populations in Premodern Xinjiang

At the beginning of the twentieth century, in Xinjiang's Uyghur villages and urban neighborhoods, daily religious life and family customs such as circumcision, marriage, and funerals were, as in the rest of sedentary societies of Central Asia, largely monitored by local religious leaders (i.e., *axun, mullah,* and *imam*). At times criticized by Westerners traveling in the area or by reformists for their cupidity, conservatism, and, sometimes, ignorance, these clerics who controlled prayer meetings in small neighborhood mosques exercised a strong authority over local communities. The larger religious sites that attracted believers to the Friday prayer or important religious celebrations were most often led by clerics who had the benefit of a longer and more thorough religious education; they came from madrassas (in Uyghur, *mädris*) where they had got a deeper knowledge of Islamic canons. Students learned how to read, write, and recite the Koran. They studied Islamic law, Persian or Arabic, and at times other sciences (i.e., Islamic history, astronomy, geography, medical science, and literature). Although the most famous madrassas were for the most part located in the Near and Middle East, as well as in Istanbul, Uyghurs were (mainly) drawn to madrassas in West Turkestan and in Tatarstan, due to cultural, linguistic, and geographic proximity.

Neighboring Islamic centers had a strong influence upon the region, especially Bukhara and Tashkent or Kazan, in the late nineteenth century. Indeed, individuals who studied in such prestigious centers[6] were considered highly influential upon their return to Xinjiang and were among the main vehicles for the spread of religious thought alongside foreign intellectuals. Though it was not systematic, at the turn of the twentieth century, many of them sought to promote a purified vision of Islam, moving away from local interpretations considered to be too conflicting with the spirit of the *sharī'a* and the orthodoxy promoted by Salafism. However, members of the official clergy were far from being intellectually unified. While some of the most conservative openly criticized heterodox practices tied to Sufism or the Cult of Saints, several were themselves members of Sufi brotherhoods (see below). There, Kashgar, Yarkand,[7] and, more generally, the western part of the Tarim Basin boasted of the most famous madrassas. The latter attracted Uyghurs coming from the Ili Valley or from Turpan, as well as members of other minority communities. Having completed their religious education in madrassas, clerics could benefit from the title of *damolla* or *çoŋ imam*, that is, superior cleric ("great molla" or "great imam"). They could also become Islamic judges (in Uyghur, *qazi*) if approved by the Qing.

Large trans-regional mystic networks (uygh. *täriqät*) interconnected with Central Asia; Afghano-Indian margins and China's Sufis networks intersected with the official clergy. Those brotherhoods were locally structured around different prayer meeting locations (uygh. *xaniqa*) and were managed by charismatic leaders (uygh. *şäyx, pir* ou *işan*). Their aura and their importance in the socio–religious life of Muslim populations of Xinjiang's oases had led them to have a strong political status especially during Khoja's theocracy[8] or Yakub Beg's[9] insurrection. During the first half of the twentieth century, the Sufi community was dominated by Nasqhbandis lineages mostly originating from western Turkestan. At the turn of the twentieth century, new branches of the Naqshbandiyya originating from western Turkestan take roots in Xinjiang. One of those powerful branches is the Naqshbandiyya-Khufiyya, known in Xinjiang under the name of Naqshbandiyya-Thaqibiyya.[10] Embedded in the spheres of influence of Islamic knowledge, its sheikhs often belonged to the official clergy and practiced a purified version of Sufism at the center of which was the mosque (mosque and *xaniqa* were conflated[11]). As illustrated in the works of Thierry Zarcone, this unbending madrassa Sufism developed in Xinjiang along with the actions of its West Turkestani sheikh founder Qamr al-Din[12] (who died in 1938) and his successors. This network quickly spread across Xinjiang.[13] Close to the deobandi naqshbandi Sufi networks, they are followers of a silent version of the *dhikr* (uygh. *zikir*) and condemn the oral *dhikr*, namely ecstatic practices that do not comply with the laws of the *sharī'a* and ishanism as well as aspects of the Cult of Saints in Xinjiang. This version of Sufism parallels more heterodox naqshbandi networks also deeply interconnected with Central Asia, the Naqshbandiyya-Jahriyya. The local branches of the Jahriyya were settled during the nineteenth century by leaders coming from Uzbekistan. Those were often tied to the Qadiriyya[14] and, in some cases, the Yasawiyya. This tainted version of Sufism revolved around Sufi prayer meeting locations in which followers engaged in ecstatic practices based upon vocal *dhikr*, music, and dance.

Ishanic practices survived outside literate Sufi circles. The masters of these circles, the Ishans (*išan*)[15] were often tied to the Jahriyya or the Islamiyya sects.[16] This hereditary Sufism was deeply ingrained in rural areas and often took the form of clientelistic relations or maraboutism. Tight relations between Sufi leaders and followers, which were based on the transmission of a deep mystical knowledge, evolved into a genuine veneration of and allegiance to leaders on the part of the local population. The latter benefited from the *barakat* that they got through the genealogy linking them to famous sheikhs or khojas (uygh. *Xoja*).[17]

Sufi Ishans and sheikhs frequently led ceremonies in the Cult of Saints. The latter was popular amongst a large proportion of Muslims in the region and well illustrated how locals had borrowed mystical elements from Islamic and pre-Islamic cultures. Though the cult was practiced across Xinjiang, it was more actively so in the areas of Kashgar and Khotan, on the tombs (uygh. *mazar* ou *gumbäz*) of local individuals to whom were granted intercession and healing powers. Some of these saints had been kings, heroes, or martyrs who had contributed to the Islamization of the region. Others were religious Sufi sheikhs or Shiite imams as in Khotan, or intellectuals, including revered craftsmen. Among the most important tombs were the mausolea of Apak Khodja,[18] Arslan Khan, Satuq Bughra Khan, and Mahmud Kashgari in the area of Kashgar, as well as those of Ordam in Yengishar, Tuyugh Ghojam in Turpan, and Imam Jäppiri Sadiq in Khotan. Those mausolea were the center of an intense religious, social, and economic life. Beyond the officers and guards looking over the mausolea, traders and shamanic healers were commonly present (uygh. *baxş* ; *büwi* for women), along with sheiks and wandering dervishes such as qalandars, who, in contrast to well-organized Sufis, would beg and lead an ascetic life, one that was similar to that of *sadhus* in India.

Islamic elites originally benefited from considerable economic power in Xinjiang. It is true that since the Qing's reconquest and the Chinese administration's move toward the full control of Xinjiang in 1884, Islamic elites' judicial authority[19] and religious endowments (uygh. *vaqf*) were reduced. However, beyond institutions built around the official clergy (i.e., mosques, *mäktäp*, madrassas, etc.), religious elites were still in control of the resources tied to the institutions of *xaniqa* and *mazar*. Simultaneously, the official clergy depended upon contributions of various sorts, the most important of which was the *ösre*. These revenues enabled them to secure their economic influence and strengthen their authority upon a still essentially agrarian society, in parallel with the Qing administration.

Jadidism and Religious Restructuration in Xinjiang

At the beginning of the twentieth century, modernity in Xinjiang did not originate from the geographically and culturally distant Chinese empire but from the Turkic world. Elites indeed resisted Qing attempts at siniciz- ing Xinjiang, including the opening of Confucian schools, and rather turned toward a model of indigenous modernization brought to light via the Russian empire and Turkey. In disagreement with the structure of traditional societies and their respective systems of representation, elites imported jadid reform- ism popular amongst Russian Tatars, during the second half of the nineteenth century. They discredited any ideas that were likely to slow down the mod- ernization of local societies or promote the assimilation of Turkic popula- tions. Fascinated by the technical and intellectual revolution that fed the rise of Western nation-states, the goal of jadids was to revitalize their culture and Turkic societies in preparation for the decolonization of the region. The dif- fusion of such ideas was based on the development of the printing industry in the region, and the spread of "scientific" schools (uygh. *pänni mäktäp*). In contrast to traditional religious schools, these institutions guaranteed a mod- ern education, notably through the teaching of Turkic literature and history, sciences, foreign languages, and sports. However, these new schools and jadid networks were subject to two forms of pressure. On the one hand, auto- cratic Chinese governors between 1911 and 1942 remained fearful that the movement could promote anticolonialism. On the other hand, traditionalist Islamic elites feared that the submission of Islam to the imperatives of mod- ernization could not only weaken its monopoly on the production of moral and religious values but also catalyze the dissolution of the order that helped secure their prerogatives.

Jadid thinking has also spread, thanks to the return of students from abroad and the west Turkestanis fleeing Stalinist repression. However, as anticolonial opposition influenced by jadidism was rising,[20] modern principles introduced by jadids slowly began to change mentalities. Emerging debates between secu- lar intellectuals and progressive Ulemas helped shape new conceptions of the role that Islam needed to play in a modern society. Educated clerics who had been trained in Bukhara or in Islamic areas of the Russian empire brought back with them new ideas shaped by jadid and Salafi reformism.[21] As in the rest of the Muslim world, the modernist project, which in Central Asia was best represented by jadidism, was accompanied by a move toward a purified form of Islam. In Xinjiang, important figures such as Abd al-Qadir Damolla (1862– 1924) played a crucial role in diffusing such syncretisms in the Kashgar area.

Secular intellectuals and reformist theologists criticized the instrumental- ization of Islam considered to be guided by dishonest interests, backward, and/or superstitious. Ishans and illiterate mullahs who exercised abusive forms of authority upon the local populations or opposed the modernization of society were particularly subject to criticisms. Similarly, superstitions and the misappropriation of resources in the context of the Cult of Saints were condemned. Moreover, the idea of "forsaking the world" supported by the Qalandars became incompatible with scientific positivism and the value of work in the eyes of jadids. Their withdrawal from society and their ecstatic practices that did not conform with the *sharīʿa* were met with the orthodox clergy's stern disapproval, as well as that of other Sufis and jadids.

Communists were similarly critical of the above religious practices,[22] especially in northern Xinjiang. During the 1930s, Soviet influence was strengthened, particularly in Urumqi and bordering areas. The new governor, Sheng Shicai, was closer to the Soviets, and ethnic minority youth were sent to study in the USSR accordingly.[23] Upon their return, several became strong defenders of secularism. Though the 1937 wave of persecutions launched by the governor led to the assassination of several religious figures and anti-communist intellectuals, the aura of communist ideology remained limited to the young intellectual circles in northern Xinjiang.

The New Socialist Society: Between State Control and Destructuration Processes

As in Soviet Central Asia, the beginning of Chinese Communist Party (CCP) power in Xinjiang in 1949 came with social and political reforms that contributed to secularizing society, transforming the range of tolerant religious activities and, as a result, changing religious life significantly.

Starting in the 1950s, the CCP's approach was to tolerate religious practices in general. Underlying this attitude was the idea that tolerance would promote the popular support the CCP needed to secure its authority in Xinjiang. In most oases, authorities still accepted Islamic education, religious activities held in mosques, as well as *xaniqa* and *mazar*, as long as religious leaders did not question the supreme authority of the party.[24]

Simultaneously, the party ensured that Islam would be subject to state control. This fear is deeply rooted in the history of Chinese sovereignty on Xinjiang. For a long time, khojas invoked Islam for the purpose of justifying the establishment of independent theocracies. During the Republican period, Islam continued to be used in southern Xinjiang by both Islamic elites and jadid militants in order to establish an anti-communist Islamic republic. In the beginning of the 1950s, Islamic elites opposed the new Socialist order, in the name of religion. As early as 1950, southern Xinjiang was characterized by severe conflicts between the Chinese authorities and religious figures as well as individuals intimately linked to "rightist" separatist networks. The most important ones were those initiated by Bardidin Makhsum[25] and sheikh Abdimit Damolla. Between 1954 and 1957, they capitalized on grievances amongst the notables (partly generated by the move toward collectivization) and mobilized through Sufi networks local population against Chinese power.[26]

Fearing that Islam would be once again instrumentalized to oppose the new policies implemented in the region, communist authorities attempted to control the local religious scene. Islamic elites' economic power slowly disappeared as a result of the communist redistribution of land. Following the law on agrarian reforms in June 1950 and the movement toward collectivization in the middle of the 1950s, mosques, *xaniqa*, and *mazar* as well as the private belongings of local religious elites were slowly collectivized. Religious institutions' ability to tax the population was prohibited. Progressively, religious elites became economically dependent upon the Chinese state, which acknowledged the most accommodative members of the clergy and sanctioned those openly opposed to the communist regime. As in the rest of the country, the clergy was placed under the authority of the Islamic Association of China (IAC, or the *Zhongguo Yisilanjiao Xiehui*).[27] The latter was founded, in abidance with the rules of the CCP, to co-opt Muslim leaders who had influence over members of the clergy and believers.

Like any political regime in the world, the new Chinese authorities have promoted through socialization institutions under their control a teleological interpretation of history serving their sociopolitical program. In Xinjiang, they focused their efforts in binding the fate of local populations with that of the Chinese nation, and in stigmatizing characters in Uyghur history, religious practices, and interpretations of Islam considered to be subversive or reactionary. Authorities judged historical figures and events on the basis of whether they were loyalists or separatists, as well as feudal or modernist. This approach to the interpretation of history led to denunciations of religious streams that had legitimated separatist or reactionary movements. Official history associated Sufism and the Cult of Saints with fanatical separatism, as well as a feudal and backward sociopolitical order. Starting in the 1980s, the publications of Chinese and Uyghur researchers drew upon this vision. Sufism and superstitions linked to the Cult of Saints were depicted as distortions of Islam that enabled khojas to maintain their authority and secure their economic domination.[28] These critiques diffused by communist institutions such as schools and the media encouraged Uyghur elites and the educated population to move away from these religious practices.

The CCP's policies increasingly became stricter. This move was in large part associated with the loss of influence of pragmatic cadres within the CCP, to the benefit of radical Maoists. The latter worked toward the eradication of religion and the short-term assimilation of national minorities. Policies of the party toward Islam and clerics in Xinjiang become stricter after the anti-rightist movement consecutive to the Hundred Flowers movement in 1956. Following a policy of tolerance in the early 1950s, the Great Leap Forward (1958–1962) and the Cultural Revolution (1966–1976) were both characterized by an overt policy of repression. Attacks targeting politicized Islam and practices considered backward left the door open to general critiques of Islam. Madrassas and religious sites were closed one after the other. Meanwhile, the move toward collectivization provoked by the Great Leap Forward fundamentally changed lifestyles. Islam was to that day excluded from the daily lives of Muslims. Despite a slight softening of religious policy following the Great Leap Forward, repression reached its peak during the Cultural Revolution. The promotion of atheism and the prohibition of Islam became the rule. Many religious sites such as mosques, xaniqas, and mazars were destroyed or transformed into warehouses or, as in some cases, into pigsties. Any person who claimed to be religious or to have played a role in the clergy was labeled counterrevolutionary. Just like intellectuals or secular cadres suspected of being local nationalists, several members of the clergy such as sheikhs, ishans, or believers suspected of disloyalty were molested, sent to reeducation camps (*laogai*), and/or even executed.

The Revival and Politicization of Islam

This period of marginalization, followed by the repression of Islam, only ended with the policy of reforms and openness (*gaige kaifang*) in the 1980s. Though ideas and activities considered threatening to Chinese sovereignty continued to be repressed, the state softened its control over society and religion. Meanwhile, this movement gave minority cadres space for greater tolerance. The freedom of religious practice was in theory guaranteed by the 1982 Constitution and the reestablishment of the once dissolved Xinjiang branch of the IAC. The softening of policies conducted by Deng Xiaoping left the door open to cultural and religious revival among the Uyghurs. That movement was accompanied by a quest for spirituality, which favored the rediscovery of Islam. The local Muslim community and, in some cases, foreign donors[29] helped rebuild mosques and construct new ones in Xinjiang. The significance of the movement in the prefecture of Kashgar reveals the growing interest in religion in southern Xinjiang at the time. Following the Cultural Revolution, the prefecture had only 392 religious sites; at the end of 1981, it had 4,700 of them, and in 1995, more than 9,600.[30] The move toward the reconstruction of religious heritage was pursued across the province until the 1990s,[31] a period during which the Chinese authorities adopted more restrictive religious policies.

The move toward re-Islamization was accompanied by the burgeoning of Koranic schools. As Chinese state control was relaxed, religious schools, often not declared, increased in number until the late 1980s when the first regulations aimed at controlling Islamic teaching were set up. Large numbers of young people from rural or clerical environments turned to the study of Islam in an attempt to embrace the career of an imam, a guarantee of social recognition at that time. According to a study in early 1990 cited by James Millward,[32] Xinjiang had 938 Koranic schools accommodating about 10,000 students. However, the movement was disorganized and anarchic. Many literate mullahs had died during the Cultural Revolution. Teaching in madrassas suffered from a lack of qualified teachers and appropriate books.[33] So, after a rudimentary initiation in their oasis of origin, talips usually attempted to join the more famed madrassas managed by clerics trained in the most prestigious Koranic schools of the republican period.

Madrassas that reopened in the Kashgar area and neighboring oases played a crucial role in the above training process. According to an official investigation quoted by Zhang Yuxi, in the four prefectures of southern Xinjiang characterized by a high concentration of Uyghurs, there were more than 665 religious classes gathering 7,081 talips (students in religion) in 1989. James Millward claims that in 1990, there were 350 such schools in the oasis of Kashgar.[34] Karghilik and Yarkand became once again major centers for religious teachings. According to Zhang, while in 1979 Karghilik had 150 talips and five Koranic schools, in 1989 the number of talips had risen to 700 and there were 33 schools.[35] In Karghilik, the rising interest in religious training was in part related to the fact that some of the schools were supervised by Mullah Ablikim Makhsum Hajji, a highly popular member of the IAC who had graduated from the prestigious *Xanliq* madrassa in Kashgar. Some of these madrassas were centers of political activism and became a source of great concern to the authorities. Zhang's interpretation of the problem was notably revealing of such preoccupations on the part of the government:

Recently [in the 1980s], a reactionary religious force emerged and developed in southern Xinjiang. [...] They use various methods to hurt patriotic religious figures, attempt to usurp the religious leadership, and cause some religious establishments to become stages for counter-revolutionary activities. [...] The seriousness of the problem is that many of the religious schools (specially the private ones) are controlled by reactionary religious powers. Some counter-revolutionaries who were released after many years of labor reform still has not changed and are spreading separatist ideas under the cover of religious teachings among young people in some Karghilik religious schools, they advertise "Jihad." [...] Moreover, many religious professionals who lead the services in Mosques have not passed the check up, creating opportunity for bad people to mix in, causing the regular religious service goes beyond its boundaries, or even causing the Mosques become the strongholds of counter-revolutionary activities.[36]

The softening of the government religious policy in the 1980s was accompanied by the politicization of Islam among the Uyghur.[37] Islam was used as a tool by both sides of Uyghur anticolonial opposition. While Uyghur nationalists were protesting that their Turk-Islamic identity was threatened by the sociocultural values promoted by the Chinese nation-state, Islamist fringes of the Uyghur opposition were promoting the setting up of an Islamic state that would put in place a social and political order "managed by Muslims and for Muslims." The number of incidents involving young talips increased, especially when the Chinese state tried to set up regulations to control Islamic teaching and shut down madrassas in Karghilik.

The most serious organization using Islam as a mobilizing tool has been the East Turkestan Islamic Party (Uygh. *Shärqiy Türkistan Islamiy Partiyisi*). The emergence of this movement is actually deeply linked to the context we have evoked about Karghilik in the mid-1980s. Zeydin Yusuf, the founder of this movement, has capitalized on the wave of discontent generated by the closure of the *madrassas* under the direction of Ablikim Makhsum in 1987–1988. Originally, the party seemed to have an anticolonial and nationalistic flavor rather than an Islamic one.[38] Testimonies from ex-members of the organization underline that the organization started expanding and getting a deeper Islamic flavor after the closure of Karghilik mosques. Actually, its founder, a high school–educated farmer from southern Xinjiang, did not have any religious background and the organization had been originally baptized the East Turkestan Liberation Party. However, the decision of the Chinese authorities to close Ablikim Makhsum's *madrassas* and to send talips back home raised strong discontent among them. Zeydin Yusuf capitalized on the talip discontent, and the general displeasure caused by their repatriation to their oases, to extend the organization networks all over eastern Turkestan. Then the organization was re-baptized East Turkestan Islamic Party and a program set up that aimed at preparing for general insurrection that would lead to the setting up of an Islamic state. It really came into prominence in April 1990 when it launched the Barin insurrection (near Kashgar). On April 5, 1990, a few dozen militants tried to enforce an ill-prepared plan of insurrection that was harshly put down by Chinese security forces. Even if the organization's members had in mind the Afghan Mujahideens' example, they did not seem to have any foreign support. However, the rise of antigovernment discourses in mosques and Koranic schools and the emergence of small jihadist groups after the dismantling of the ETIP[39] have fed government concerns[40] during the 1990s.

The Tightening of State Control over
Religious Activities

While the conservative wing was taking over CCP management following 1989, the Chinese state was also tightening its control over society and religious life in Xinjiang and attempted to stabilize the region using a carrot and stick policy. The carrot policy was to promote economic development in Xinjiang[41] and the stick policy sought to sanction severely or eliminate any elements or forces considered to be potentially subversive.

State control, which deepened throughout the 1990s, became more systematic following the Strike Hard campaign, launched in 1996. In Xinjiang, the campaign targeted ethnic separatism and antigovernmental Islam. Strategies revolved around the need to rid the CCP and local administrations of unreliable elements: to strengthen propaganda against separatism and religious extremism, to reinforce state control over indigenous populations, to encourage the flow of cadres and Han Chinese into Xinjiang in the context of the Xinjiang Production and Construction Corps, to severely limit the construction of new mosques, to grant leadership positions in mosques or religious organizations to those who love the nation, to register and monitor all people trained in an unregistered religious school, and to take strong measures against the penetration of Islam into social and political life.[42]

The set of policies aimed at controlling Islam was founded upon the judicialization (chin. *fazhihua*) of religious activities, a new approach to the management of religion that moved away from state policies of earlier decades. Indeed, until that period of time, there were no extensive sets of laws regulating religious affairs. The degree to which the state controlled religion was a direct function of Politburo guidelines. Starting in 1988, the government of the Xinjiang Uyghur Autonomous Region (XUAR) established temporary regulations on the administration of religious sites. In October 1990, six months after the events in Barin, the government passed temporary regulations on the administration of religious activities and personnel. In July 1994, religious activities were managed on the basis of a new series of guidelines: the Administration of Religious Affairs regulations in XUAR,[43] and a new and stricter national regulation defined in 2001 and then promulgated in 2004.[44]

In order to be legal, today, a religious activity has to be tied to one of the five official religions in China,[45] have state-designated clergy as its organizers, take place in government-registered sites, and respect CCP ideology. Starting in 2001, with the intention of pursuing its struggle against antigovernmental Islam, Beijing launched a campaign aiming for the patriotic reeducation of imams. The latter, who had already been under tight surveillance by the local Religious Affairs Bureau and the IAC, were from then on obliged to take patriotic education classes in order to re-frame their religious discourse and rectify any potential deviant behavior. The tightening of control and the sanctions that came with it helped eliminate antigovernment discourse across mosques. Furthermore, the behavior of individuals whose career depended upon the state administration changed. Hence, several CCP members, bureaucrats, teachers, and students limited their religious involvement fearing that it would have detrimental effects upon their professional life.

As far as Islamic education is concerned, the new legislation prohibits unregistered Koranic schools. Without proper authorization, imams are not allowed to provide religious teaching including during the *mäshräp*.[46] Doing so would increase their risk of being punished. To be officially recognized by the authorities, young imams have to graduate from Islamic institutes (in Chinese, *Yisilanjiao jingxuexiao,* and in Uyghur, *islam dini inistituti*).[47] The three-year curriculum managed by the IAC aimed to train members of the clergy who conform with the main lines of the CCP. Simultaneously, the Chinese authorities closely monitored the translation and publication of Islamic works, the activities of visiting foreigners, as well as Uyghurs' overseas stays, in order to avoid the penetration of subversive foreign religious influences into Xinjiang.

Foreign Islamic Influences in Xinjiang

In spite of these measures, Uyghurs have reconnected with the rest of the Islamic world. During the 1980s, the progressive opening of borders enabled them to circulate more freely and provided opportunities for spiritual exchanges and revival, especially among the youth. This led to an increasing interest in religious ideas originating from Uyghurs living abroad or Muslim visitors in Xinjiang. Before the advent of the Internet, Muslim visitors would stealthily carry religious publications in their suitcases into Xinjiang. Their books would then be translated into Uyghur, copied, and carried under their vests. At that time, traders, preachers, students, and relatives coming from abroad began to proselytize, taking advantage of the relaxing of Chinese policy. Pakistani traders who would head to southern Xinjiang and to Urumqi were, following the opening of the Karakoram Highway, particularly influential amongst the Uyghurs who for decades had felt isolated from the *Ummah*. Meanwhile, an increasing number of Uyghurs, traders, and students traveled to the Muslim world. Wealthy and some middle class religious Uyghurs bypassed the strict regulations limiting the number of pilgrims authorized to go for the Hajj. Some youth also took advantage of the networks that grew out of the increase in the flow of people studying Islam abroad.[48] Hence, until the mid-1990s, hundreds of young Uyghurs (according to official sources, there were thousands of them) attended religious schools in Pakistan,[49] Egypt,[50] Turkey,[51] and

Saudi Arabia. A few of them also headed to Yemen, Qatar, and Malaysia for the same purpose. Those who returned to Xinjiang[52] were often influenced by the deobandi, salafi, or wahabbi interpretations to which they had been exposed throughout their religious training.

The influence of these Islamic currents has spread to neighboring regions as well, albeit marginally, thanks to the neo-fundamentalists networks operating in Pakistan and Central Asia. Some of them began to preach discretely in order to avoid Chinese authorities' strict controls over Islam in Xinjiang. A few preachers of the Pakistani *Tabligh-e-Jamaat* proselytized in southern Xinjiang and Urumqi during the 1990s. However, the expansion of *Tabligh* was impeded by the linguistic and cultural barriers separating the Uyghurs from the Pakistanis, as well as by the Chinese and Pakistani governments' strengthening of their controls on the flows of Pakistanis and Uyghurs in and out of Xinjiang. In recent years, Uyghurs coming back from western Turkestan have established a few cells of the *Hizb-ut Tahrir* in Xinjiang. The organization is far from being as deep rooted as it is in neighboring Central Asia. However, the potential reach of this underground network's vehement criticism of CCP policies made Chinese authorities pay special attention to this phenomenon. In that respect, in 2007, the government launched a campaign with the specific purpose of uncovering and eradicating *Hizb-ut Tahrir's* cells in the area of Kashgar and Yarkand.[53]

Toward the Spread of Purified Versions of Islam?

Sufism and Ishanism have been deeply impacted by the Cultural Revolution. It has led to a decline in religious knowledge by affecting or breaking knowledge transmission lines. It is true that the beginning of the 1980s came with the revival of Sufi networks and the Cult of Saints. However, interviews conducted in the course of the last decade suggest that these practices seem to have declined since the 1990s, especially among the educated youth. The diffusion and penetration of foreign influences, combined with the tightening of state control and co-optation strategies, favor the evolution toward more orthodox religious practices. As highlighted earlier, a certain prestige is associated with foreign interpretations of Islam among some circles of Uyghur youth. Among them the aura of salafi currents is strong and the tenets of these currents of Islamic thought often criticize heterodox practices such as the Cult of Saints or some Sufi practices.

Beyond these foreign influences, state policies also had significant effects. The critique of Sufism and Ishanism that pervaded the educational system, the media, as well as literature contributed to diverting elites and educated youth. Moreover, the retrograde and separatists dimensions associated to the

Khodjas' hold on traditional society in the eyes of the Chinese authorities often led the state to co-opt non-Sufi clerics generally considered "saner" and more progressive. The interruption of Islamic teaching, the dismantling of madrassas, and the arrest and the death of certain clerics during the Cultural Revolution impacted strongly on local Islamic practices. However, clerics influenced by early twentieth-century Central Asian reformism who refrained from confronting the Chinese state were able to promote again their orthodox interpretation of Islam—one that is less impacted by practices such as heterodox Sufism or cult of the Saints.

Finally, Sufism and popular religious practices, such as the Cult of Saints, suffered from the judicialization of religious activities: the new legislation implemented between 1990 and 2010 aimed to render illegal any activity that does not conform with the strict guidelines established by the law. The law tends to limit religious activities to the spectrum of orthodox Islamic practices. As a result, a large part of Sufi activities are considered illegitimate, as numerous sheikhs in the region were not officially recognized as Islamic clerics by the government. In short, for most zealous local officials, acceptable Islamic practices beyond their pro-governmental involvement tend to be limited to orthodox practices taking place in the mosques of some townships. Hence, even if Sufi practices and the Cult of Saints[54] are not explicitly prohibited, their lack of legal recognition has made them more vulnerable to repression and prohibition during periods of conflict with or crackdown by the government.[55]

Conclusion: Uyghur Islam Caught between Restructuring and Confrontations

In 1949, the diffusion of jadid reformism, the spread of communist ideology, and the implementation of CCP authority over Xinjiang fundamentally transformed interpretations and manifestations of Islam as well as Muslims' relation to it.

Hence, for a large number of believers, especially urban elites in northern Xinjiang, the existent connection between religious spaces found in mosques, Sufi lodges (i.e., *xaniq*), and *mazars* slowly began to crumble. By closely monitoring Islamic practice and education in Xinjiang, and by attempting to eliminate customs considered backward or potentially subversive, reformist jadids, ulemas, and the Chinese authorities contributed to strengthening the foundations of more orthodox versions of Islam. In that sense, the elimination of "feudalism" based on the restructuring of traditional systems of representation and the eradication of religious elites' power, as in the former Soviet Union, led to a relative decline in Sufism, Ishanism, and the Cult of Saints.

While the latter continued to be practiced in southern Xinjiang, especially in rural areas, in the 1980s, the diffusion of proselytizing interpretations of Islam from neighboring Pakistan, Central Asia, and the Arabic world favored the rise of purified interpretations of religion, especially among the urban population.

Beyond this evolution in religious interpretation and practice, the twentieth century was characterized by the marginalization of Islam and its elites in Uyghur society. The latter phenomenon was partly related to modern elites and the communist state's willingness to avoid the use of Islam as a tool of political contestation against the broader campaign for the social transformation of Chinese society. The practice of religion thus became confined to the private sphere, the educational system was secularized, and Islamic elites who once benefited from considerable economic power began to depend on Chinese state recognition in order to get emoluments.

Furthermore, in contrast to state intentions during the Cultural Revolution, today, the objective of the central government is not to attack Islam per se. Rather, its intentions are to avoid the religious legitimation of a separatist or antigovernmental discourse in Xinjiang.[56] In attempting to do so, the CCP keeps in mind the Hui as a model. Although the Hui have maintained their religious and cultural specificities, communist authorities have succeeded in transforming historically stormy relations with the Hui into peaceful ones. This was notably facilitated by the alliance between the government and the *Yihewani* current, a powerful branch of the Muslim brotherhood in China.

However, as far as Uyghurs are concerned, the problem is qualitatively different and more complex. Even if the Hui are drawn toward the Muslim world in many ways, cultural proximity with the Chinese world, a better socioprofessional insertion, and wider cultural and religious freedoms favor the feeling of being plenipotentiary members of the Chinese nation. Uyghurs seem to face a completely reversed situation. Several factors have impeded their potential to identify with the Chinese nation. As far as the CCP is concerned, the Chinese nation was born out of a desire to "live together," based on a common history and an adherence to communist egalitarian ideals. However, Uyghur national identity mobilizes a specific Turkic ancestrality, culture, and prestige.

The feeling of having inherited powerful empires,[57] which at times were rivals of China, and past and recent independentist experiences of the republic of East Turkestan[58] arouse doubt as to the idea of a shared history and the inevitability of the integration of Xinjiang with the rest of China.[59] However, the dangers of assimilation caused by the Chinese model of modernization and the persistence of economic inequalities across ethnic lines in Xinjiang, in light of the tight control of political institutions, are interpreted as a manifestation of colonialism, hence discrediting Beijing's enthusiastic promise of equality and social justice during the Maoist times. In other words, the process whereby Uyghurs are increasingly confronted with the presence of a Han Chinese authority in control of the economy and local administration in Xinjiang has both accentuated the feeling of being second-class citizens and increased antipathy to the colonial rule. Hence, on the fringes of Uyghur society that fear sinicization through the model of modernization promoted by the Chinese state (see below), Islam is both a source for an alternative sociopolitical model and a tool for political mobilization.

Due to a lack of supportive religious elites, the CCP has not been able to rely upon the Uyghur clergy as much as it has on the Hui clerics to defuse ethnic conflict. The tight monitoring of the clergy, religious sites, and religious education centers since 1990 seems to have successfully eradicated antigovernmental discourses, which were commonly encountered in mosques and underground madrassas. Nonetheless, these policies of tight control have generated frustrations among the Uyghurs. Today, the obligation of the clergy to represent the CCP lines prevents it from being an intermediary between believers and the state, as it may have been the case among the Hui. Hence, the new role of imams as spokespeople of the government and their lack of room for maneuver has led some of them to ignore rising frustrations among the Uyghur community and has thus in some cases discredited them. This phenomena makes Uyghur clergymen fear that they could be bypassed by underground militant groups capitalizing on the Uyghurs' frustrations if they loose their role of middlemen between Chinese administration and Uyghur believers.

Notes

1. Rémi Castets has worked as a research fellow at the French Centre for Research on Contemporary China (CEFC, Hong Kong) between 2004 and 2006. Since 2006, he is lecturer at the University Michel de Montaigne Bordeaux III where he teaches geopolitics and Chinese political history. He is also junior research associate at CERI-Sciences Po (Centre for International Studies and Research, Paris).
2. Both modern "Uyghurs" and "Uzbeks" are populations resulting from the turcization of an old Iranian ethnic fund.
3. This toponym was taken back by the anticolonial Jadid reformists and the anticommunist modern Uyghur political opposition. By using this toponym, those circles assumed the region called "Xinjiang" ("new border") since its conquest by the Qing empire was a predominantly Turkic territory.
4. See infra.
5. Here, we define modernity as the multiple recompositions in cultural, social, political paradigms following the development of science in the West. As emphasized by authors such as Paul Ricoeur, those deep changes stand on the development of new systems of interpretation of the world diffusing through mass communications. Those new systems either stood as an alternative to traditional systems of interpretation greatly inspired by religion or reinterpreted them in order to make them more compatible with the sociopolitical order they promoted. Moving away from the idea that human beings were subjects of a greater

divine order, individuals began to question traditional cosmogonies and see themselves as the architects of society and the world through the development of modern ideologies. On those phenomena, see, for instance, P. Ricoeur, *Anthologie* (works selected and presented by Michaël Foessel and Fabien Lamouche) (Paris: Seuil, 2006), pp. 369–370.

6. These centers included Bukhara, Samarkhand, and, starting at the end of the nineteenth century, Tashkent and Kazan.

7. Testimonies from the older generation of clerics refer to some of the most famous madrassas, including the *Xanliq* and *Qazançä* madrassas in Kashgar as well as the *Çoŋ Mädris* in Yarkand.

8. Kachgaria was ruled at that time by a naqshbandi theocracy who took power over the declining local gengiskhanid houses. On this naqshbandi theocracy and its links with Central Asia sufi networks, see A. Papas, *Soufisme et politique entre Chine, Tibet et Turkestan* (Paris: Maisonneuve, 2005).

9. Just like naqshbandi khojas, he also relied on Sufi networks to impose his authority upon the region in the 1860s and 1870s.

10. For information about the Naqshbandiyya networks in Xinjiang during the twentieth century, see T. Zarcone, "Sufi Lineages and Saint Veneration in 20th-Century Eastern Turkestan and Contemporary Xinjiang," in H. Celal Güzel, C. Cem Oğuz Osamn Karatay, *The Turks,* vol. 6 (Ankara: Yeni Türkiye Publications, 2002), pp. 534–537.

11. It is worth noting that the conflation of official clergy and Sufi networks was frequent in that particular area.

12. Fleeing the Ferghana Valley in 1926, he settled down in Karghilik, and then in Yarkand (see T. Zarcone, "Sufi Lineages and Saint Veneration," p. 535).

13. The opening of *Çoŋ Mädris* in Yarkand in 1945 by his successor, Ayub Qari, and the reputation of the institution favored the expansion of that network.

14. In the region, Jahriyya networks share common genealogies with Qadiriyya and Yasawiyya networks. Those networks, especially the Yasawiyya, were deeply rooted in western Turkistan a few centuries ago.

15. The latter term was used by some Sufi leaders to invoke their genealogical ties with prestigious sheikhs, especially with naqshbandi khojas.

16. The Islamiyya networks mentioned by the specialists of Sufism in the region seem to trace their origins to old local branches of the Jahriyya in the region (T. Zarcone, "Sufi Lineages and Saint Veneration," pp. 537–538).

17. Religious leaders would gain their legitimacy out of their genealogical history linking them to the Prophet, important Sufi saints, as well as Genghis Khan.

18. He was the founder of a naqshbandi theocracy that dominated Uyghur oases before the Qing conquest and was revered by both Turkic-speaking and Chinese-speaking Muslims for being one of the main initiators of Islam and Sufism in Northwestern China.

19. Its purposes had been strictly reduced to matrimonial conflict resolution.

20. These anticolonial nationalist militants played a key role in the short-lived Turkish Islamic Republic of East Turkestan (TIRET). Managed by the emirs of Khotan and anti-communist jadids, the latter Republic was based in the areas of Khotan and Kashgar and lasted between November 1933 and February 1934.

21. On the linkages between jadidism and salafism in Central Asia, see S. Dudoignon, "Djadidisme, Mirasisme, Islamisme," *Cahiers du monde russe,* no. 1–2 (1995), pp. 13–40. Also refer to T. Zarcone, "Un aspect de la polémique autour du soufisme dans le monde tatar, au début du XXᵉ siècle: Mysticisme et confrérisme chez Mûsâ Djarallâh Bîgî" in S. Dudoignon, D. Is'haqov, R. Möhämmätshin (eds.), *L'Islam de Russie* (Paris: Maisonneuve et Larose-IFEAC, 1996), pp. 227–248.

22. See, for instance, Shähidi, Burhan, *Şinjang 50 yilim* (Beijing: Millätlair Naişriyati, 1986); Saifudin, *Saifudin huiyi lu* (Beijing: Huaxia chubanshe, 1993).

23. A. Abdurahman, *Taşkäntçilär* [with the following subtitles in Mandarin: "Those Who Studied in Tashkent"] (Urumqi: Şinjang Xälk Näşriyati, 2002).

24. Interviews conducted in Xinjiang and amongst exiled communities between 1998 and 2007.

25. Bardidin Makhsum was close to the anticommunist exiled leader Mehmet Emin Bughra.

26. Abdimit Damolla was a follower of the powerful Ayub Qari. Based in Yarkand, the latter sheikh led the Naqshbandiya-Thaqibiyya in Xinjiang. As an anti-communist, he had connections with Mehmet Emin Bughra and expressed his opposition to the Chinese authorities shortly before his death in 1952, in circumstances that remain unclear to this day. A few months later, Abdimit Damolla and Bardidin Makhsum launched an organization called *Salam*, whose purpose was to prepare the upheaval of East Turkestan and the reestablishment of an Islamic regime.

27. Established in 1953.

28. With respect to this critique, see T. Zarcone, "Le culte des saints de 1949 à nos jours," *Journal de l'Histoire du Soufisme*, vol. 3 (2001), pp. 160–164.

29. Especially rich Saudi Arabians.

30. See G. Bovingdon, "Autonomy in Xinjiang: Han Nationalists Imperative and Uyghur Discontent," *Policy Studies*, vol. 11 (2004), p. 33.

31. At the turn of the year 2000, the region had about 24,000 mosques, namely two-thirds of mosques in China.

32. See J. Millward, *Eurasian Crossroads. A History of Xinjiang* (New York: Columbia University Press, 2007), pp. 324–325.

33. Interviews, 1998–2007.

34. J. Millward, *Eurasian Crossroads. A History of Xinjiang*.

35. Y. Zhang, "Xinjiang jiefang yilai fandui minzu fenliezhuyi de douzheng ji qi lishi jingyanti," in F. Yang, Li Ze, and Dong Sheng (eds.) *Fan yisilanzhuyi, fan tujuezhuyi yanjiu lunwenji* [*Research on Pan-Turkism and Pan-Islamism*] (Urumqi: Xinjiang shehui kexueyuan, 1994). These numbers have been slightly underestimated by ancient *talip* from Karghilik who are now settled abroad (Interviews in 2006).

36. Y. Zhang, "Xinjiang jiefang yilai fandui minzu fenliezhuyi de douzheng ji qi lishi jingyanti."

37. The phenomenon was accentuated by already existent grievances among the Uyghurs in regards to their unprivileged social and political position.

38. Interviews, 2003–2007.

39. In the early nineties, former sympathizers of ETIP set up small networks known as the East Turkestan Reformist Party (ETRP) and the East Turkestan Islamic Party of Allah. However the most famous of Uyghur underground jihadist movements is known in the West as the East Turkestan Islamic Movement (uygh. *Şärqiy Türkistan Islamiy Härkäti / Şärqiy Türkistan Islamiy Partiyisi*). Close to the first two networks mentioned above, its core members decided to break Uyghur Islamist militants' isolation. His leader, Hassan Makhsum, a sympathizer of the ETRP, decided in the late 1990s to move to Pakistan and Afghanistan to escape the constant pressure of Chinese security forces in Xinjiang and set up a sanctuarized organization that could send back trained militants to Xinjiang to prepare its insurrection. Using connections with the Taliban, cadres from the Islamic Movement of Uzbekistan (IMO) who retreated to northern Afghanistan, and Al Qaeda networks, this movement has trained dozens of militants. However, the East Turkestan Islamic Movement was disorganized and weakened by the operation the Western coalition has led in Afghanistan after 2001.

40. Before their dismantling, these organizations destroyed and attacked military bases and organized terrorist incidents (i.e., the assassination of officials, Han Chinese, Uyghur dignitaries and bureaucrats, as well as bomb attacks). On the political forms taken by Uyghur opposition since the 1980s, see R. Castets (2004), "Nationalisme, Islam et opposition politique chez les Ouïghours du Xinjiang," *Les Etudes du CERI*, vol. 110 (2004), <http://www.ceri-sciencespo.com/publica/etude/etude110.pdf> (accessed March 3, 2010).

41. This policy has been promoted since 2000 by the program aiming to develop the western region of China (*xibu dakaifa*).

42. Those guidelines were exposed by the Political Bureau of the CCP Committee during a special meeting organized in March 1996 on the stabilization of Xinjiang. See: "Guanyu weihu Xinjiang wending de huiyi jiyao, zhongyang zhengzhiju weiyuan hui," March 19, 1996, <www.taklamakan.org/guidebook/Doc7.htm> (accessed March 3, 2010).

43. These regulations result from the promulgation of a new legislation on the control of religious activities.

44. See Human Rights Watch, "Devastating Blows. Religious Repression of Uighurs in Xinjiang," *HRW Special Report*, vol. 17, no. 2 (2005). For a more official version of the legislation see, Ma, Pinyan, "Dang de zongjiao zhengce zai Xinjiang de shijian," *Xinjiang shehui kexue*, no. 1 (2005), pp. 49–55.

45. These are Buddhism, Daoism, Catholicism, Protestantism, and Islam.

46. The mäshräp are gatherings at the local level, favoring the communication of traditional Uyghur culture.

47. The first institute in the region opened in 1987 in Urumqi. The one in Kashgar opened in 1991 and then was expanded in 2002. During that latter year, it accommodated 150 students.

48. Some of them went to inner China to study in Hui Koranic schools, where restrictions were lower.

49. Some Chinese scholars have suggested that 10,000 Uyghurs left China to study in Pakistan (*International Herald Tribune*, October 15, 2001). Among them, several were students of the Islamic University of Islamabad. Starting in the end of the 1990s, Pakistani authorities began to deport talip with an irregular status and pressured local Koranic schools not to accept students originating from the other side of the Pamirs.

50. Especially at Al-Azhar University. Today, in order to be admitted to the latter, Uyghur students require approval from the local Chinese Embassy.

51. The University of Marmara admitted Uyghur students until 2000–2001, when it decided to reduce their number. Marmara alumni who stayed in Istanbul then opened the Association for East Turkestan Culture and Solidarity in 2006 (in Turkish, *Doğu Türkistan Maarif ve Dayanışma Derneği*). This association seeks to promote, among other things, a 'purified' interpretation of Islam among the Uyghurs.

52. However, students who have left Xinjiang without authorization are now subject to increasingly harsh state punishment upon their return. This has created incentives on the part of overseas Uyghurs not to go back to Xinjiang once they have completed their studies.

53. See "Kashidiqu jizhong kaizhan jiepi 'Yisilan jiefang dang' fandong benzhi xuanjiao huodong," *Xinjiang Pingan Wang*, June 8, 2007, <http://www.xj.xinhuanet.com/pingan/2007–06/08/content_10251470.htm> (accessed March 3, 2010).

54. To this day, the authorities have limited collective gatherings around large *mazar*.

55. See for instance "Xinjiang: How Long Will Arrested Sufis Be Held," *Forum 18 News Service*, September 26, 2005, <http://www.forum18.org/Archive.php?article_id=659&pdf=Y> (accessed March 3, 2010).

56. Ma, Pinyan, "Xinjiang fandui feifa zongjiao huodong yanjiu," *Xinjiang Shehui Kexue*, no. 4 (2003), pp. 72–78; Li, Jiansheng, "Guangyu yindao Xinjiang yisilanjiao yu shehui zhuyi shihui xiang shiyin de wenti," *Xiyu Yanjiu*, no. 1 (2001), pp. 72–78.

57. Huns, Kaghanate (744–840), Karakhoja Uyghur Kingdom (856–1335), Karakhanid Empire (940–1212).

58. Including the founding of a Khodja theocracy in the seventeenth century, the Emirate of Yakub Beg (1865–1878), the Turkish Islamic Republic of East Turkestan (TIRET) (1932–1933), as well as the East Turkestan Republic (1944–1949).

59. On the official historiography of the region and its populations, see for instance "History and Development of Xinjiang," *White Paper from the State Council of the People's Republic of China*, March 2003, <http://www.china.org.cn/e-white/20030526/index.htm> (accessed March 3, 2010).

CHAPTER 15

The Jama'at al Tabligh in Central Asia—a Mediator in the Recreation of Islamic Relations with the Indian Subcontinent

BAYRAM BALCI[1]

It is well-known that post-Soviet Central Asia—Kazakhstan, Kyrgyzstan, Uzbekistan, Tajikistan, and Turkmenistan—is going through a process of re-Islamization, driven by both local dynamics and foreign influences from the Middle East and Turkey. The Islam of the Indian subcontinent, although lesser known and not as powerful an influence, is nevertheless contributing significantly to this re-Islamization. It merits attention all the more as it represents a continuation of a rich relationship of exchange with the Indian subcontinent that has spanned centuries.

This chapter offers an exposition of how this re-Islamization process takes place, with a focus on the missionary activities of two major Indian Islamic organizations: the *Ahmadiyya* and, more significantly, the *Jama'at al Tabligh*. How they are perceived in the varying contexts of Kazakhstan, Kyrgyzstan, and Tajikistan will be discussed, based on fieldwork conducted in these countries. We will also address how they organize action from New Delhi, Deoband, and Lucknow, based on interviews of Central Asian students and activists in these centers. Although this recent resurgence is part of a long tradition of religious exchange between India and Central Asia, our purpose here is to analyze more specifically the features and dynamics of these missionary movements with the aim of integrating them into the broader picture of Central Asia's re-Islamization, and to assess their tangible impact on local communities and societies.

How Ancient Are Islamic Exchanges between Central Asia and India ?

Although predated by Buddhism,[2] Islam is among the oldest traditions that have strengthened relations between India and Central Asia. In India, Islamization dates back to the Arab conquest but it was under the Ghaznavid Empire (962–1187) and the Delhi Sultanate (1206–1526) that it was given a strong impetus.[3] It then reached its apex under the Mughal Dynasty, founded by Babur, a descendant of the Timurid Dynasty that originated in the Ferghana Valley of present-day Uzbekistan.

During these periods, the main carrier of the Islamic faith and values abroad was the world-renowned Sufi brotherhood of the *Naqshbandiya*, founded and established in the town of Bukhara by Abdukhalik Ghijduvani, Bahauddin Nakshiband and Khwaja Ahrar. Subsequently, however, Babur and his mentor Baqibullah (1563–1603), who followed him in his military expeditions to India,[4] had greater success with dissemination. Later, another key figure of the Nakshibandi community, Ahmad Sirhindi (1564–1624) played no less a significant role in the resurgence of a somewhat rejuvenated form of the brotherhood in Central Asia.[5]

All religious ties were summarily broken in the second half of the nineteenth century, when tsarist Russia invaded and subjugated Central Asia. The succeeding Soviet regime proved even more suspicious of religious subversion and hermetically sealed all Soviet borders to prevent foreign religious, particularly Muslim, influence. Although the Soviet authorities always kept a very close watch on religious groups, there were nevertheless limited but continuous exchanges with religious centers in India. Thus, during Soviet times, Domla Hindustani (1895–1986), a great figure of Central Asian Islam, did visit India and received religious education in Deoband—an education that he passed on to numerous disciples. These disciples later initiated the Islamic revival that followed the collapse of the Soviet regime and the establishment of Central Asian republics as independent nations in the 1990s.

Islamic Revival in the Newly Independent States and Renewed Religious Links with the Outside World

The collapse of the Soviet Union and the unexpected emergence of independent states in Central Asia impacted the redefinition of local Islam, which had been weakened by decades of Soviet religious repression. Religion was not banished any more, and traditional as well as emerging elite classes encouraged Islamic revival, viewing it as a major component of and contributor to national identity and state legitimacy. Nevertheless, at the same time the regimes remained suspicious about the potential for foreign religious influence to develop out of control,[6] especially when originating from the Middle East, Turkey, India, and Pakistan. The latter two turned out to be quite influential, thanks to the efforts of three movements: the earlier mentioned *Naqshbandiya*, the transnational *Ahmadiyya*, and last but not least, the *Jama'at al Tabligh*, which began disseminating in Central Asia for the first time in history.

Since the establishment of the independent Central Asian states, many recent studies have focused on the local revival of Sufism,[7] which flourished under the ambivalent attitude toward Islam in Uzbekistan. Yet President Karimov has vacillated between supporting and promoting traditional Islamic values as a source of tolerance and social peace, and suspecting and even condemning it for being far too strong a mobilizing force for political protest.[8] In the early 1990s, President Karimov's religious policy, in close cooperation with Turkey, which still had excellent diplomatic relations with Uzbekistan at the time, focused on the rehabilitation of the Nakshibandi mausoleum outside Bukhara. It was developed into a huge complex with the objective of making it a center of pilgrimage for thousands of Uzbek and foreign pilgrims. This open policy helped develop and tighten renewed links with foreign Nakshibandi branches, especially with organizations based in Turkey, where the Nakshibandi legacy is still very much alive. However, it was a Nakshibandi sheikh originating from Pakistan who became immensely successful and popular in the local and transnational community of Nakshibandi sufis. Sheikh Muhamad Zulfikar Nakshibandi Mujaddiid from Lahore first came to Tajikistan and Uzbekistan where he initiated young disciples into Sufism. Among them was Salim Buhari, a well-educated Uzbek who had graduated in the field of German Studies, and who in 2008 became the director of the Bahaduddin Nakshiband foundation in Bukhara. Soon, Salim Buhari developed contacts with Nakshibandis in Pakistan and became influential in his community. His increasing popularity and influence were soon perceived by official power as a potential threat to the stability of the regime. Therefore, the Uzbek authorities tried to limit his activities and have been rejecting his exit visa applications for the past four years.[9] Nevertheless, his influence among the Sufi communities of Central Asia is substantial, especially in Tajikistan and Uzbekistan, although the full extent of it is difficult to estimate.

The second India-based movement to develop activities in Central Asia is the Ahmadiyya, the Sufi community founded by Mirza Ghulam Ahmad (1838–1908), who was often referred to as a prophet.[10] He used to introduce himself as "the Mahdi"—the Messiah—and believed he was a reincarnation of Jesus Christ, although this did not inhibit him from borrowing Hindu symbolism such as the image of Krishna. His movement thrived at a time when Indian Muslims were experiencing heavy pressure from foreign Protestant missionaries who accused them of being inherently violent, while the Hindu sect of Arya Samaj was active and successful in converting Muslims into Hindus.[11] In this context of threats from various directions, the Ahmadiyya developed rapidly to an international scale, mobilizing forces for what was perceived as a necessary defense of Islam.[12]

Similar to the actions taken by the Jama'at al Tabligh to defend and promote Islam, the Ahmadiyya implemented innovative missionary methods derived from their rivals, the foreign Protestants colonizing India. At the time, Mirza Ghulam Ahmad recommended that "classical" jihad using force and armed violence should be given up and replaced with "metaphoric" jihad—called *bil lisan*—that would be pure rhetorical preaching based on the principles of nonviolence.[13]

However, the movement rapidly split into two branches, though with quite blurred ideological boundaries. The dominant of the two, the Qadani (named after Qadian, the Indian native town of Mirza Ghulam Ahmad) lays great emphasis on the prophetic status of Hazrat Mirza and his successors, who are venerated as caliphs. In contrast, the lesser of the two branches, called the Lahori (from Lahore), holds that Hazrat Mirza was a *mujaddid*, a reformer, and not a new prophet; this latter belief would set the Sufis and the rest of the Umma at odds. Both branches of the Ahmadiyya have experienced severe repression within the Muslim community, especially in Pakistan[14] after the partition, and in Saudi Arabia, where they were accused of schismatic behavior and were forbidden to access the holy places of Mecca and Medina.

Although they are discredited and repressed by the Muslim elites, the disciples of the Ahmadiyya have succeeded in spreading their message worldwide to Africa, Europe, and North America. In the former socialist block they have made remarkable inroads in the extremely secular and notoriously isolated Albania.[15] In Central Asia, they have exerted great efforts to infiltrate into the various states of the region, but so far only Kyrgyzstan, where religious freedom is stronger and political power less suspicious of foreign religious proselytism, has proved open to the teachings of the Ahmadiyya.[16] Even though the Ahmadiyya face hostility and harsh criticism from other Muslim schools, they have succeeded in being legally registered in Kyrgyzstan under the Ministry of Justice and under the State Commission for Religious Affairs (SCRA). In the course of our investigations, we have searched for them in the other Central Asian states, but they seem to be absent. Further study will be needed, however, as they are known to operate very discreetly, a tactic they were forced to develop to escape repression.

The Influential Jama'at al Tabligh

Till the present day, the Jama'at al Tabligh is the most influential of all India-based Islamic movements active in Central Asia. It was founded in the 1920s in the remote state of Mewat by Muhammad Ilyas Kandhalawi (1885–1944), who was a Sufi sheikh of great renown throughout the length and breadth of British India, even though scholars still disagree on the genuine Sufi character of his teachings.[17] The Arabic term *tabligh* means "to deliver the message" and indeed the overall objective of the Jama'at is to encourage Muslims, and Muslims only, to revive their faith and religious practice, for they hold it to be one of their first and foremost duties. To this end, the *Tablighis*, as they are called, strictly obey Islamic law and dogma regarding dress, rituals, and traditions. So far, their efforts exclusively target Muslims, with the aim of raising their spiritual consciousness. They prefer to think of their movement and activities as non-political, a position that has historical roots, as their founder Muhammad Ilyas encouraged his followers to remain on the sidelines of the volatile political differences between Muslim and Hindu communities in pre-partition India.

At first, the Jama'at concentrated their work exclusively on strengthening Muslim faith and practices in India itself, where Islam had in many cases been diluted by pre-Islamic elements. Later, the Tablighi missionaries began to spread their word to the rest of the world—including France, where since the 1960s the association *Foi et Pratique* (Faith and Practice) has provided religious support to north African immigrants in low-income suburbs.[18] The Tablighis are amateur preachers who travel the world on their own private funds, spreading the message that the central rule of *Tabligh* must be strictly obeyed. They are not professional "clerics," but rather volunteers and family men coming from a broad spectrum of social and professional backgrounds. Before departure, they are specifically trained in religious seminars, whose length and content are adapted according to individual profiles and needs; these can range from three-days workshops to sessions lasting 10, 20, or 40 days, with some courses taking even four months.[19] Candidates have to go through a step-by-step preparation before they are eligible for the title of *qadim,* or "experienced senior." This is the point of departure from all other Muslim proselytizing organizations. In the Jama'at, the *da'wa*—the mission—is not exclusively reserved to a limited circle of educated religious men and "clerics."[20]

Indeed, the Jama'at al Tabligh has developed a new kind of mission. In the times of the Prophet and his successors, the state was responsible for the *da'wa*, which was closely related to the *jihad,* so as to disseminate Islam. Both *da'wa* and *jihad* were seen as a collective community duty. Muhammad Ilyas Kandhalawi radically reformed the concept itself and made *jihad* a nonviolent individual duty for every single Muslim, who must dedicate some time and energy on a day-to-day basis for spreading the good word in order to deserve the title of "good Muslim."[21]

According to some members of the Jama'at who were interviewed, the first Tablighi missionaries made their way to Central Asia in the Soviet times as early as the 1960s and 1970s, thanks to student exchange programs with India

that were quite developed at the time.[22] One can reasonably doubt this assertion, but what is certain is that the Jama'at made scant difference here until the
collapse of the USSR and the subsequent independence of all Central Asian
republics in the early 1990s. Though the Tablighis are eager to show that they
have long been active in the region, in actuality they started to develop only
after 1991, when these countries opened up to a large variety of foreign religious influences. The Tablighis seem to struggle more than others in adapting
to Central Asian contexts, where the most influential movements originate
from the Arab Middle East or from Turkey, the latter having the advantage of
sharing a common ethnolinguistic background with Central Asia.[23]

Although it is difficult to precisely date the establishment of the Jama'at al
Tabligh in the region, it is highly probable that their interest in Central Asia
intensified after the 9/11 attacks, even though paradoxically this was when the
regimes of all Central Asian states increased their already tight and repressive
control over all radical or political Islamic organizations. The Tablighis are highly
visible, as they dress in salwar kameez and engage in door-to-door preaching, yet
they were never apprehensive of repression. As the Jama'at introduces itself as a
nonpolitical, nonradical movement of pietists, they seem to have benefited from
the beginning from the favorable consideration of the authorities. The latter
seem to believe that the Jama'at al Tabligh provides an alleviative and harmless
alternative for disorientated youth, who are easy preys of radical Islamic groups.
Still, in the general context of official repression of most foreign Islamic movements in Central Asia, the relatively lenient treatment of the Jama'at puzzles
analysts but reflects ignorance of the authorities regarding this group.

The Jama'at al Tabligh is not equally active throughout all of Central Asia.
At present, it is totally absent from Turkmenistan and only slightly active in
Uzbekistan and Tajikistan. It is in Kazakhstan and Kyrgyzstan that they are
the most successful.

Uzbek authorities, in their attempt to totally control all aspects of
Islam—including the training of imams, the publication of literature, and
dissemination of day-to-day religious practices and rituals—officially
refused to give legal status to the Jama'at al Tabligh. The rare disciples
who have tried in the past years to spread the word were arrested and
given heavy jail sentences, even though the very charismatic former Uzbek
mufti Muhammad Sadik Muhammad Yusuf said that he considered the
Jama'at as an inoffensive and nonpolitical movement.[24] Many people were
arrested and jailed for having organized allegedly Tablighi meetings. The
regime has been successful in fighting the spread of radical Islam, and all
movements that are not legally registered under the Ministry of Justice are
considered to belong to radical Islam, such as the Islamic Movement of
Uzbekistan[25] or Hizb ul Tahrir,[26] both of which are under tight control and
experience harsh repression.

In Tajikistan, despite the relative tolerance of the government toward independent Islamic organizations in general—thanks to the Islamic Renaissance Party's representation in the coalition government—the Jama'at al Tabligh has no legal status. Indeed, official attitudes toward Islamic groups range from tolerance to repression. In August 2009, during our latest fieldwork in Dushanbe, several Tablighis had just been arrested and accused of membership in an illegal Islamist organization. The ban on the Jama'at al Tabligh is, paradoxically, attributable to the Islamic Renaissance Party, as it certainly perceives the Jama'at as a potential competitor, while the secular components in the present Tajik government consider the Tablighis as harmless nonpolitical pietists.[27]

In Kazakhstan, the Jama'at al Tabligh is tolerated. The Tablighis organize meetings and sermons, but they can face problems at any time as they do not have any legal status. Authorities refuse to grant them registration under the Ministry of Justice and under the Muftiate (Muslim Board) on the grounds that they may use it for politicizing their activities. However, despite this sword of Damocles hanging above their heads, the Tablighis in Kazakhstan have so far never been bothered by the police.[28]

Thus far the Jama'at al Tabligh has met with its greatest success in Kyrgyzstan, where it was granted an official and legal status. We found Tablighis not only in the capital Bishkek, but also in Naryn and in the south in the Ferghana Valley towns of Osh, Jalal Abad, and Batken, where religious consciousness and practice are stronger than anywhere else in the country. Several factors in addition to greater religious freedom explain this relative success. The particular and local nature of Islam favored the importation of the "minimal Islam" preached by the Tablighi *dawachis*[29]—those who preach the *da'wa*. It is true that Kyrgyz Islam is "light"—because of a late and weak Islamization of the population—and tinged with shamanistic elements. The Tablighi dawachis, being amateur preachers with no high religious education, focus on back-to-basics actions targeting disorientated youth so as to educate them in the essential rules of Islam and turn them away from deviant behaviors such as alcoholism and drug addiction. They teach them how to pray and read the Koran and encourage them to get involved in preaching and to share their faith. More than any other Central Asian country, Kyrgyzstan has proved to be a fertile field of development for the Jama'at.

Moreover, in the Osh region where we interviewed many Tablighis, we found that most were ethnic Kyrgyzs. Even though ethnic Uzbeks account for 50 percent of the regional population, they are barely present in the movement. This may be due to the fact that the Jama'at recruits more successfully among the Kyrgyzs, whose faith is rather lukewarm, while the more devout Uzbeks are less receptive. The fact that ethnic Kyrgyz are in general poorer and have looser community and social ties is also part of the explanation.

Wherever they go, the Tablighi dawachis implement the same and unchanging methods of preaching and spreading their faith. Trained as they were in

India, Pakistan, and sometimes even as far away as Bangladesh or Central Asia by missionaries originating from these countries in the 1990s, the Tablighis obey the rules of proselytism as defined by the Jama'at's founders in the 1920s. As mentioned above, they regard peaceful jihad as an obligation for every Muslim, who is obliged to dedicate some time and energy to this fundamental duty. Thus, for example, after Friday prayer, they circulate in small groups doing door-to-door proselytism and inviting people to attend religious meetings at the mosque. In Bishkek, we took part in one of these *gasht* (from the Persian "go on patrol") groups, which also serve to recruit new members, who then go onto forming such new groups under the supervision of a senior in the community. These new groups go on patrol and in turn bring new disciples to be initiated in the Tabligh. Similarly, newcomers are required to form small groups that go on training patrols lasting from 3 to 40 days. They circulate around the villages and must demonstrate their ability to recruit and persuade people to join a gasht.

The community leaders are not easy to identify, as the movement considers itself as egalitarian and without hierarchical structure,[30] yet every single member is required to strictly obey the rules of gasht and da'wa at least three days each month, and for 40 consecutive days each year. Meetings usually take place in the neighboring mosques, as the movement does not possess any building or office. Returning from gasht, the disciples are invited to meet under the leadership of an *amir*, to whom they report on the latest gasht campaign, that is, on the number of homes they visited, and on the answers and opinions they collected. Then, the amir gives a long sermon named *beyan,* which is usually a commentary on a Koranic verse or the hadiths. Moussa Khedimallah, who has been studying the Jama'at's presence in French suburbs, observed similar missionary practices, showing that their activities in Central Asia are no different from their standard methodology in other countries.[31]

The Jama'at al Tabligh versus Central Asian Islam

It is not an easy task to measure the level and nature of connections between the Jama'at al Tabligh and other Islamic groups and institutions, as disciples of different persuasions barely communicate with or acknowledge each other. All such groups must initially face two major institutions controlling Islam: the State Commission for Religious Affairs (SCRA) and the Muslim Board for Spiritual Affairs (Muftiate), which is supposedly an independent organization but a state-sponsored one, and whose leader is the general mufti.[32] In all Central Asian countries, the Jama'at al Tabligh has fairly strained relations with these institutions, which do not share their vision and objectives, and which hardly understand what their purpose is. Further, the Tablighis' appearance, particularly their Pakistani salwar kameez and long beards, exasperates officials who themselves exert great efforts to appear as Western as possible. For example, in Kazahstan, in July 2009, we had the opportunity to interview the mufti of the central mosque in Almaty, who acknowledged that he had no idea of the identity or provenance of the individuals wearing salwar kameez and sporting beards and coming to the mosque at every single prayer every day, right behind him. He admitted that it had never occurred to him to inquire who they were and, in fact, asked us to inform him about the Jama'at and the dawachis, and their objectives and methods.

Official imams appear to feel helpless against these generally young and uneducated grassroots activists who succeed in filling mosques with new devotees. Officials are puzzled and do not know how to react, as the Tablighis are successful and at the same time respect the law and never do anything morally objectionable or against the society. In Kyrgyzstan, the movement is all the more welcome since the Muslim Board has a specific department for da'wa that is in charge of supervising the spreading of Islam in the country. And yet, da'wa is considered legitimate only so long as it is organized under the strict supervision of the state. Therefore, the organization of gasht, its itinerary of villages and other places, the funds allocated to it, and all the preaching rules are negotiated at the highest level with the religious state. This goes as far as defining a dress code, although authorities have no means to control who wears what, and the fact remains that the Tablighis still dress in the Pakistani fashion.[33]

The Jama'at al Tabligh is not the only independent Islamic organization involved in missionary activities; there are others as well, although they operate without any legal status or official approval. Such is the case with the Hizb ul Tahrir.[34] Though illegal, they are said to be present in the Ferghana Valley. Its overall objective is openly political as it campaigns for the restoration of the caliphate to reunify the world's Muslim community, the Umma, into a single religious state. It is on these grounds that they were banned from all Central Asian countries, especially from Uzbekistan, where they were relatively successful till 2000 but have subsequently experienced harsh repression. The Hizb ul Tahrir has hardly ever had connections with the Jama'at al Tabligh, although this assertion is difficult to prove. By virtue of its structure and ideology it is at the very opposite end of the spectrum from the egalitarian, nonpolitical, and nonradical Jama'at. When interviewed, the Tablighis expressed their disagreement with the methods of the Hizb ul Tahrir; indeed, they regard the activities of the Hizb ul Tahrir as a reprehensible *fitna*—a source of dissension among the Muslim community.

In the power struggle for influence over Islam in Central Asia, the Jama'at al Tabligh faces far more powerful rivals, such as Turkish Islamic groups that benefit from the good diplomatic relations between Turkey and Central Asian states as well as their common ethnolinguistic backgrounds. First, the disciples of the Nakshibandi sheikh Osman Nuri Topbas opened several small madrassas in Central Asia.[35] His works were translated into vernacular languages and Russian and are well disseminated across the region, especially in Kyrgyzstan and Kazakhstan. Another Turkish organization known as the *Suleymanci*, named after their leader Suleyman Tunahan (1881–1959), controls several Koranic schools, which are in operation despite their lack of legal status.[36] Ultimately, the most influential organization is a branch of the Turkish neo-brotherhood of the Nurcu, founded by Said Nursi (1876–1960), also known as the Fethullahci, named after Nursi's disciple and charismatic leader Fethullah Gulen (born in 1938).[37] Their community is divided but they are successful in developing covert methods of proselytism. All these Turkish groups have never been in direct confrontation with the Jama'at al Tabligh, as they do not target the same audience. Indeed, Turkish movements appeal more to educated Muslims and devotees and focus their action on educational cooperation or trade between Turkey and Central Asia. The prevailing view is that the various missionary groups do not know each other, do not communicate with each other, and remain unaware of the others' activities. In Azerbaijan, where similar re-Islamization is in progress and involves influential foreign groups, the Muslim Board organizes coordination meetings in order to initiate dialogue. Central Asia lacks this type of initiative.

The question of the actual connections between the Tablighi networks in Central Asia and the Indian subcontinent is critical to this discussion. As mentioned above, the Jama'at initially infiltrated Central Asia through the efforts of committed missionaries from India, Pakistan, and Bangladesh who arrived in two successive waves in the early 1990s and around 2001. Now, however, these volunteers have stopped coming in. Fieldwork conducted in October 2009 and February 2010 in New Delhi, Deoband, and Lucknow shows a reversal of direction, with exchanges now relying on Central Asians visiting the high places of the Jama'at in India.

In the heart of the traditional district of Nizamuddin in New Delhi, the historical headquarters of the Jama'at al Tabligh are a welcome place for Tablighi travelers from all over the world, including Central Asia. Although it does not give the outward appearance of being organized for this purpose, visitors are invited to stay at the center and attend training that can last from 3 to 40 days and sometimes even 4 months. Every night, during the usual *beyan* (sermon), a local or guest Islamic authority preaches in Urdu or in Arabic, and his speech is translated into Arabic, Russian, or English according to the composition of the audience. When we attended, a Moroccan French disciple was translating into French for a small group of Francophones with Arabic, Turkish, and French backgrounds.

On each of our visits to the center, we found groups of ten to twelve "pilgrims" from the former Soviet republics—young people from Kazakhstan, Kyrgyzstan,

Tajikistan, and Chechnia, Tatars from Russia, and even a couple of Uzbeks from Kokand listening to the sermon in Urdu with the help of a Tajik disciple, who had been staying at the center for a long time. As these converted native Central Asians come of their own accord to India to be initiated in the Tabligh, the Jama'at can rely on them to spread their message home. After stays of varying duration in the center at Nizamuddin, visitors proceed with their journey, visiting other high places of the Jama'at in India, that is, Deoband and Lucknow, as well as Bangladesh. The director of Dar al Ulum, the biggest madrassa in Deoband, explains that after 9/11, Indian authorities no longer granted long-term student visas to foreigners with the exception of a small number of Malaysians and Indonesians.[38] Thus the Tablighi converts from Central Asia stay only for short periods of time. They are all always welcome to the community home of Deoband, dar al dhuyuf, while attending short training sessions.

In contrast to other international Islamic groups and transnational networks, the Jama'at al Tabligh is not a platform for visitors to develop business connections. The visits of the young disciples from Central Asia we met were all dedicated to religion, with the exception of the Uzbeks, who are unable to obtain any Indian visa other than a business visa. The few Uzbeks we met had resorted to the importation of car parts that are unavailable in Uzbekistan as their pretext for obtaining a visa that would enable them to come and visit the Tablighi places of India. In their particular case, they succeeded in integrating their professional activity with their faith. They are taken in charge by the visitors' Tablighi center at Nizamuddin where accommodation is free and where they seek to raise their spiritual consciousness. In doing so, they take a risk, since the Jama'at al Tabligh is forbidden in Uzbekistan. Should the Uzbek government discover that Uzbek nationals are being initiated in the Tabligh in India, these individuals might be pursued and charged with illegal religious activity. When asked about their motivations versus the risks, one of them, whose name cannot be disclosed here, said "We are aware that we are taking risks, but we have to do so, since otherwise we cannot do anything anymore. A few years ago, we were allowed to meet legally and talk about religion, whereas now we are under so much pressure, that it has became impossible because we are afraid of what might happen to us. The only way to escape this fear and strengthen our faith is to seek help abroad, especially here in Delhi, where brothers from all over the world give us hope and strength."

The Role of Islam in Balancing Geopolitics between the Indian Subcontinent and Central Asia

Until the tsarist Russian invasion of the nineteenth century, longstanding mutual influences had created close religious ties between Central Asia and

India, especially under the Mughal Dynasty. Although the Soviet Union maintained good relations with India, it ended up disrupting these ties. In the vacuum of ideology and identity left by the collapse of the Soviet system, many foreign religious movements became involved in an intense competition. Among the Indian and Pakistani groups, the Jama'at al Tabligh is definitely the most successful in terms of the number of missionaries who came to spread the word as well as their contribution in raising religious consciousness among undereducated and disorientated youth, especially in the relatively poor nation of Kyrgyzstan. Their success can be measured as well by the relative number of converted Tablighis who visit India to get initiated and trained before returning home to continue proselytism.

However, even though their influence is increasing, it is still very weak in comparison to the achievements and influence of other Islamic groups based in Turkey or the Middle East, as can be seen from the number of Central Asian youth studying in Egypt, Syria, and Turkey. Moreover, their influence is limited to marginalized youth, with virtually no impact on the elite classes, suggesting that for the time being they are not pursuing political goals and that they are having no effect on diplomatic and political relations.

In the Central Asian context, the Indian and Pakistani organizations that are active in Central Asia must so far be considered merely as transnational faith-based movements. The Tablighis have been successful in so far as they addressed the spiritual needs of an ever-increasing number of individuals among the poor and the young. But they carry no political weight and claim to pursue nonpolitical goals. Nevertheless, the spreading of their word throughout the whole of Central Asia could easily encourage them to widen the scope of their action and audience to reach the elite classes, especially given the deteriorating social and economic conditions and the lack of religious and political freedoms that are mobilizing masses and empowering radical groups with alternative projects.

Notes

1. Director of the French Institute for Central Asian Studies (IFEAC), www.ifeac.org, email: balci_bayram@yahoo.fr

2. G. Fussman, É. Ollivier, B. Murad, *Monuments bouddhiques de la région de Kaboul / Kabul Buddhist Monuments*, vol. II (Paris: Publications de l'Institut de Civilisation Indienne du Collège de France, fasc. 761 et 762, Paris, 2008). Voir également P. Leriche, S. Pidaev, *Termez sur Oxus Cité capitale d'Asie centrale* (Paris: Maisonneuve et Larose-IFEAC, 2008).

3. About the first contacts between India and Islam, see: M. Gaborieau, *Un autre islam, Inde, Pakistan, Bangladesh* (Paris: Albin Michel, 2007). D. Matringe, *Un islam non arabe, horizons indiens et pakistanais* (Paris: Editions Téraèdre, 2005).

4. See the rich and informative memoirs of Babur: *Babur-Nama, mémoires du premier Grand Moghol des Indes, le Livre de Babur* (Paris: Imprimerie nationale, 1986).

5. J.-A. Gross, "The Naqshbandiya Connection: From Central Asia to India and Back (16th–19th Centuries)," in S. Levy (ed.), *India and Central Asia. Commerce and Culture, 1500–1800* (Oxford: Oxford University Press, 2007), pp. 233–259.

6. M. B. Olcott, D. Ziyaeva "Islam in Uzbekistan: Religious Education and State Ideology," Carnegie Papers, no. 91, July 2008, <www.carnegiendowment.org> (accessed March 3, 2010).

7. T. Zarcone, "Ahmad Yasavî héros des nouvelles républiques centrasiatiques," *Revue du monde musulman et de la Méditerranée*, no. 89–90, 2000, pp. 297–323.

8. M. B. Olcott, "Sufism in Central Asia. A Force of Moderation or a Cause of Politicization?" *Carnegie Papers*, n° 84, May 2007.

9. M. B. Olcott, *Sufism in Central Asia*.

10. C. Servan-Schreiber, "Le mouvement Ahmadiyya: Une organisation musulmane trans-nationale et missionnaire originale," *Après-Demain*, Journal mensuel de documentation politique, *L'islam dans le monde*, n° 447–449, October–December 2002, pp. 8–12.

11. About this sect, considered as a reformers' movement, see Britannica Encyclopedia: <http://www.britannica.com/EBchecked/topic/37454/Arya-Samaj> (accessed March 3, 2010).

12. About the Ahmadiyya, see Y. Friedmann, *Prophecy Continuous: Aspects of Ahmadi Religious Thought and Its Medieval Background* (Berkeley: University of California Press, 1989).

13. Ibid.

14. C. H. Kennedy, "Towards the Definition of a Muslim in an Islamic State: The Case of the Ahmadiyya in Pakistan," in D. Vajpeyi, Y. Malik, *Religious and Ethnic Minority Politics in South India* (Riverdale, MD; London: Jaya Publishers, 1989), pp. 71–99.

15. N. Clayer, "La Ahmadiyya lahori et la réforme de l'islam albanais dans l'entre-deux-guerres," in V. Bouillier, C. Servan-Schreiber, *De l'Arabie à l'Himalaya. Chemins croisés en hommage à Marc Gaborieau* (Paris: Maisonneuve et Larose, 2004), pp. 211–228.

16. About the official attitude toward the Ahmadiyya in Kyrgyzstan, see <http://www.forum18.org/Archive.php?article_id=322> (accessed September 11, 2009).

17. A polemical question over which analysts disagree. Even the contemporary disciples of the movement disagree on the connections binding the Jama'at with sufism. See: M. GABORIEAU, "What Is Left of Sufism in Tablîghî Jamâ'at?" *Archives de sciences sociales des religions*, vol. 135 (2006), <http://assr.revues.org/index3731.html> (accessed September 11, 2009).

18. About the Jama'at al Tabligh in France, see: M. Khedimallah, "Esthetics and Poetics dimensions of apostolic Islam in France: The Emblematic Case of Young Preachers of the Tabligh Movement," *Isim Newsletter*, 2003, p. 20.

19. B. Metcalf, *Islamic Revival in Brithis India: Deoband, 1860–1900* (Princeton: Oxford University Press, 2002); M. K. Masud, *Travellers in Faith: Studies of the Tablighi Jama'at as a Transnational Islamic Movement for Faith Renewal* (Leiden: Brill, 2000).

20. M. Ilyas, *A Call to Muslims* (Lyallpur, India: Malik Brothers, 1944).

21. About the different forms and notions of proselytism, and the key notions of da'wa, tabligh, jihad in particular, see: M. K. Masud, "Ideology and Legitimacy," in M. K. Masud, *Travellers in Faith: Studies of the Tablighi Jama'at as a Transnational Islamic Movement for Faith Renewal* (Leiden: Brill, 2000), pp. 79–121.

22. R. C. Horn, *Soviet-Indian Relations, Issues and Influences* (New York: Praeger, 1982).

23. B. Balci, "Between Da'wa and Mission: Turkish Missionary Movements in Central Asia and the Caucasus," in Rosalind Hacket (ed.), *Proselytization Revisited: Rights Talk, Free Markets and Culture Wars* (London: Equinox Publishers, 2007), pp. 367–390.

24. I. Rotar, "Uzbekistan: Why Were Some Tabligh Members Given Lesser Jail Terms than Others?" Forum 18, December 3, 2003, <http://www.forum18.org/Archive.php?article_id=468&pdf=Y> (accessed September 15, 2009).

25. The Islamic Movement of Uzbekistan is the most radical Islamist and jihadist organization in Uzbekistan. Founded at the end of the soviet era on the ruins of several Islamic organiza-tions that had failed to gather people in the Ferghana Valley in the beginning of the 1990s, it became a powerful paramilitary organization and a threat to the regime. Based in Tajikistan

until the end of the civil war in 1997, they benefited from favorable circumstances before moving to Afghanistan and developing relations with the Taliban and Al Qaeda. Their ties with the latter proved fatal to the Islamic Movement of Uzbekistan, as American bombings on Afghanistan in November 2001 killed Juman Namangani, the military commander of the movement. Tahir Yoldashev, the political leader of the party survived but failed to revive the movement that is believed to struggle for surviving in the remote region of Waziristan. About the history and development of the Islamic Movement of Uzbekistan, see D. Chaudet, "Terrorisme islamiste en Grande Asie centrale: Al Qaïdisation du djihadisme ouzbek," *Russie. Nei. Visions* no. 35, December 2008.

26. The Hizb ul Tahrir is an Islamic party that emerged in Jordan in the 1950s and thrived in Central Asia in the 1990s. Their utopist goal is to gather Muslims all around the world for the restoration of the Caliphate. Though they do not appeal to armed action, their speech is very hostile and radically against the regimes of Central Asia, which in return took harsh repressive action against the Hizb ul Tahrir. See D. Chaudet, "Hizb ut-Tahrir: An Islamist Threat to Central Asia?" *Journal of Muslim Minority Affairs*, vol. 26, no. 1 (2006).

27. Interview with Said Ahmad Kalandar, Tajik researcher, Dushanbe, July 2009.

28. I. Rotar, "Kazakhstan: Punished for Preaching in Mosques," *Forum 18*, November 14, 2006, <http://www.forum18.org/Archive.php?article_id=868> (accessed September 15, 2009).

29. M. Toktogulova, "Le rôle de la *da'wa* dans la réislamisation au Kirghizistan," *Cahiers d'Asie centrale* [En ligne], vol. 15/16 (2007), <http://asiecentrale.revues.org/index77.html> (accessed December 15, 2009).

30. D. Reetz, "The 'Faith Bureaucracy' of the Tablighi Jama'at. An Insight into Their System of Self-organization (Intizam)," in Gwilym Beckerlegge (ed.), *Colonialism, Modernity, and Religious Identities: Religious Reform Movements in South Asia* (Oxford, New Delhi: Oxford University Press, 2008), pp. 98–124.

31. M. Khedimallah, "Les jeunes prédicateurs du mouvement Tabligh, la dignité identitaire retrouvée par le puritanisme religieux?" *Revue anthropologique*, no. 10 (2001), pp. 5–18.

32. *Central Asia: Islam and the State*, International Crisis Group, ICG Report no. 159, <http://www.crisisgroup.org/library/documents/report_archive/A401046_10072003.pdf> (accessed March 3, 2010).

33. Interview with Ravshan Eratov, head officer for the Predication department in the Board for religious affairs of Kyrgyzstan, July 2009.

34. See also note 25. Scientific works on the Hizb ul Tahrir in Central Asia are rarely satisfying, as the object escapes the curiosity surrounding its actions and keeps a low-profile so as to hide from justice and repression in countries where it operates without legal status. However a couple of studies can be referred to: On the Party in general and its ideology, see: S. Taji Farouki, *A Fundamental Quest: Hizb al Tahrir and the Search for the Islamic Caliphate* (London: Grey Seal, 1996). On the Party's activities in Central Asia, see: Emmanuel Karagiannis, "Political Islam and Social Movement Theory: The Case of Hizb-ut Tahrir in Kyrgyzstan," *Religion, State and Society*, vol. 33, no. 2, 2005, pp. 137–149.

35. Osman Nuri Topbas's ideas are displayed at length online in French (www.terredepaix.com), in Turkish (www.altinoluk.com, the monthly journal of the movement), and in a community website (www.gonuldunyamiz.com).

36. A Suleymanci leader that we met in July 2009 in Bishkek claims they are managing 19 schools in Kyrgyzstan and 42 in Kazakhstan. Their size varies from twenty to three hundred students at a time. The website www.tunahan.org provides a good insight into Tunahan's ideas and achievements.

37. B. Balci, *Les missionnaires de l'islam en Asie Centrale, les écoles turques de Fethullah Gülen* (Paris: Maisonneuve et Larose-IFEAC, 2003).

38. Interview with the Rector of the Dar ul 'Uluum madrasa, Deoband, October 2009.

INDEX

Abkhazia, 11, 14, 34

Abu Yahia Al Libi, 37

Afghanistan, 1–6, 12, 20–21, 25–27,
29–31, 35–38, 40, 49, 61–93,
112, 124–125, 127–128, 132–136,
139, 150–151, 156, 158–163, 166,
168–169, 173–194, 202, 232, 248
See also Af-Pak (strategy)

Af-Pak (strategy), 4, 62, 72, 74, 76,
86–88, 91, 127–128, 141, 150, 173,
189, 192
See also Afghanistan

Ahmadiyya (movement), 235, 237–238, 247

Aini (base), 21

Akbar, 200–204, 207, 209, 212

AktobeMunayGaz, 158

Almaty, 26, 28–29, 32, 149, 159,
168–169, 243
See also Kazakhstan

Al-Qaeda, 4

Amu-Darya (basin), 158

Andijan (riots), 31, (city), 148, 197, 199

Arcelor Mittal, 164

Arunachal Pradesh, 98–99, 103, 111,
113–114

Asia-Europe Meeting (ASEM), 16

Asian Development Bank, 111,
161–162, 186

Asia-Pacific (region), 51

Association of Southeast Asian Nations
(ASEAN), 30, 100, 102, 121

Atyrau, 33, 157, 159, 162

Atyrau-Alashankou (pipeline), 33

Aynak (copper field), 67, 184

Azerbaijan, Baku, 123, 141–145, 156, 244

Babur, 197, 199–205, 207, 210–211, 236, 246

Bagtiyarlyk (field), 158

Baku-Tbilisi-Ceyhan (oil pipeline),
141–148, 152

Baku-Tbilisi-Erzurum (gas pipeline),
141–143, 145, 147–148
See also Caucasus

Baluchistan, 69, 90, 151, 205

Bandar Abbas (port), 152, 163

Bangladesh, 100, 242, 244–245

Bengal (bay of), 17

Berdymukhammedov, Gurbanguly, 37

Bhutto, Zulfikar Ali, 65, 88

BRIC (countries), 10, 100

Buddhism, 5, 99, 233, 236

Caravan (trade), 209

Caspian (Sea), 2, 6, 20, 33, 36, 124,
141–143, 146–148, 150–152,
(region), 69, 139, 142–143,
145–148, 150–151, 156–159, 162

Caucasus, 12, 16, 143, 145
See also Azerbaijan, Baku

Central Asia-South Asia (CASA),
161–162, 189

Central Asia-South Asia Regional
Electricity Market (CASAREM), 189

Central-South Asia Transport and Trade
Forum (CSATTF), 162, 186

Chahbahar (port), 70, 163, 180

Chechnya, Chechen, 14, 29, 65, 122, 143
China National Nuclear Corporation
     (CNNC), 165
China National Offshore Oil Corporation
     (CNOOC), 33, 109, 157
China National Petroleum Corporation
     (CNPC), 19, 33, 36–37, 108, 110,
     115, 149, 157–158
China's Metallurgical Group, 90
China's People's Liberation Army (PLA),
     27, 31–32, 34, 37, 38, 120
Chinese Communist Party (CCP), 35,
     220–221, 225–230, 233
Chinese Development Bank, 164
Chinese National Corporation for
     Heavy Machinery (CHMC), 164
Cold War 9, 29–30, 41–44, 61, 77, 105,
     121, 210
Collective Security Treaty Organization
     (CSTO), 10, 19, 21, 34
"Color" revolutions, 19, 31, 206
Commonwealth of Independent States
     (CIS), 13, 34, 163, 165, 190

Dalai Lama, 5, 98–99, 102–103, 113
Darkhan (field), 158
Deng, Xiaoping, 222
Deoband, Dar al Ulum, 235–236,
     244–245
Dharamsala, 102–103
Dostyk-Alashankou (border post), 157, 160
Drug trafficking, narcotics, 3, 18,
     29–30, 47, 62, 64–65, 78, 124, 128,
     174–176, 178, 184, 186, 188
Durand (line), 65, 85, 87, 91

East Turkestan Independence Movement
     (ETIM), 31
Economic Cooperation Organization
     (ECO), 64, 68, 77, 185, 186
Egypt, 145, 226, 246
Ekibastuz, 161
Electricity, 12, 19–20, 108, 115, 123,
     125–126, 149, 160–162, 164, 180,
     186, 189
Erzurum, 141, 143, 145

Eurasia, Eurasian, 2, 10, 13, 29, 34, 82, 87,
     89, 142–144, 148, 151, 174, 190, 192
Eurasian Economy Community (ECC), 10
European Union, 1–3, 9, 11, 18, 38, 72,
     117, 136, 155–156, 165, European
     Commission, 144, 178, 186, 188

Fakhrabad (polygon), 20
Far East, 3, 9, 15–16, 19, 22–23, 151
Faryab (province), 181, 188
Ferghana (valley), 34, 197, 199, 236,
     241, 243, 247

Gail, 157, 159
Gazprom, 17, 19
Georgia, Tbilisi, 31, 34, 141–143, 145
Gold mines, 35
"Good neighborhood" (policy), 1, 28, 30
"Great Game," 3, 56, 133, 142, 210
     See also *Heartland* (notion)
Guangdong Nuclear Power Group
     (CGNPC), 164–165
Gwadar (port), 5, 89, 100, 103, 148, 152

Hazaras, 67
Heartland (notion), 1, 51, 123, 130
     See also "Great Game"
Highways, 176, 188
Himalayas, 98
Hindustani, Domla, 236
Hizb-ut Tahrir, 227, 240, 243, 248
Hormuz (strait of), 89, 103–104
Huawei, 66, 108

Ili (valley), 217
Indian (ocean), 17, 89, 101, 103, 113,
     123, 162
Indian Oil Corporation (ICO), 157
Indian Space Research Organization
     (ISRO), 17, 167
Indian Technical Economic
     Cooperation Program (ITEC), 168
Information technology (IT), 17, 55,
     106, 121, 127, 135, 150, 166–168
International Atomic Energy Agency
     (IAEA), 17, 101, 114, 165

International Security Assistance Force (ISAF), 64, 66–67, 178, 179, 190, 191
Inter-Services Intelligence (ISI), 73
Iran, 1–3, 9, 13, 17, 31, 37, 63, 65, 67–70, 72, 74–77, 81, 84, 109, 113, 125, 127–128, 136, 143, 145, 149–152, 158–163, 169, 175, 182–183, 186, 188–192, 197–199, 202, 205, 215
Iran-Pakistan-India (pipeline), 150–152, 159
Iraq, 31, 84, 86, 145, 150
Islam, 5–6, 21, 36, 61, 128, 200–201, 207–208, 212–213, 215–233, 235–246
Islamic Association of China (IAC), 221–223, 225–226
Islamic Development Bank, 161
Islamic Movement of Uzbekistan (IMU), 37, 232, 240, 247–248
Islamic Renaissance Party, 241
Islamism, Islamist, 3, 5, 11–13, 18–20, 25, 31, 37, 47, 49, 65–66, 72, 76–77, 104, 123, 127, 132, 224, 232, 240–243, 245–248
Ittipak (Kyrgyz Uyghur society), 36

Jadidism, Jadid, 219–220, 228, 230–231
Jalalabad, 90, 181, 194
Jammu and Kashmir, 49, 69, 75, 128, 160, 190
   See also Kashmir
Japan, 1, 3, 15–16, 23, 30, 36, 38, 99, 101, 104–106, 108–109, 115, 130, 136, 149, 151, 156–157, 180, 186, 188
Jiang, Zemin, 28, 29, 39
Jihad, 62, 69, 73, 223–224, 232, 238–239, 242, 247

Kadeer, Rebiya, 36
Kandahar, 163, 177, 180–182, 188–189, 191, 205, 211
Karakoram Highway, 85, 90, 226
Karimov, Islam, 63, 237
Karshi-Khanabad (base), 31
Karzai, Hamid, 62, 68–69, 74, 87–88, 90, 174

Kashgar, 148, 152, 160, 205, 217–219, 222–224, 227, 231, 233
Kashmir, 3–6, 16, 49, 65, 69, 75, 81, 85, 87, 90, 104, 112–113, 128, 132, 148, 160, 190, 205
   See also Jammu and Kashmir
Kazakhaltyn, 163
Kazakhstan, 3–4, 10–11, 13, 19, 21, 26–28, 30–33, 36, 39, 64, 109, 123–126, 134, 142, 145–152, 155–169, 186, 235, 240–241, 244, 248
   See also Almaty
Kazatomprom, 164
KazMunayGas, 157–159
Kenkiyak (field), 157–158
Khandahar, 90
Khorgos (border post), 36, 158, 160
Khotan, 218, 231
Kunduz, 63, 161, 171
Kurmangazy (field), 150, 158
Kyrgyzstan, 3–4, 11–13, 21, 26, 30–36, 39, 63–64, 123, 155–156, 159–161, 164–165, 186, 189, 235, 238, 240–241, 243–244, 246–248

Lavrov, Sergey, 65–66
"Look North" (policy), 41–43, 45–55
Loya Jirgha, 174
Lukoil, 19

Mackinder, Sir Halford, 1
Madrassa, 216–218, 222–224, 228, 239–241, 244–245
Maharashtra, 17
Manas (base), 31, 36
Manchuria, 34
MangistauMunayGas, 158
Mao Zedong, 30
Mazar-e-Sharif, 162–163, 171, 181, 188
McMahon (line), 98, 103, 113
Military exercise, 15, 31–32, 34
Military-industrial complex, 12, 15, 17–18, 20
Mongolia, 26, 31, 39, 104, 136, 161
Muftiate, 241–242

Mughal (dynasty), 197–207, 209, 211–213, 236, 246
Muhammad Sadik Muhammad Yusuf, 240
Mujahideen, 49, 224
Multilateralism, 28–29, 53
Mumbai, 4, 162

Nabucco, 141–146, 148, 151–152
Naqshbandi (sufi order), 207, 217, 231–232, 236–237, 244
Narasimha Rao, P.V., 41
National Association of Software and Service Companies (NASSCOM), 168
Nazarbaev, Nursultan, 168
Nehru, Jawaharlal, 45, 98, 100, 102, 110–111, 120
Niyazov, Saparmurat, 36
Nizamuddin, 244–245
Nonalignment, 16, 43, 45, 102
Non Proliferation Treaty (NPT), 16–17, 103
North Atlantic Treaty Organization (NATO), 14, 29, 62–64, 66–67, 73, 76, 87, 91, 143, 179, 191–192
North Caspian Sea Project, 36
Northern Alliance, 49, 67, 70
Northern Distribution Network, 64, 163
North-South (corridor), 17, 150, 152, 161–163, 189–190
Nuclear energy, 101, 125–126
Nurcu (neo-brotherhood), 244

Obama, Barack, 62, 72–74, 76, 86, 90, 173
Oil and Natural Gas Corporation (ONGC), 17, 108–110, 115, 157–159
Organization of the Islamic Conference (OIC), 64, 68, 77
Osh, 159–160, 241

Pakistan, 1, 3–6, 16, 20–22, 24–25, 31, 35, 37–38, 45, 49, 62–63, 65–70, 72–77, 81, 83–92, 98, 100, 102–104, 112–113, 125, 127–128, 129, 132–137, 148–152, 156, 159, 161–162, 169,

173–175, 178, 183, 186–193, 226–227, 229, 232–233, 237, 242, 244
Pamir, 156
Parvez Kayani, Ashfaq, 72
Parwar (province), 66
Pashtun, 65, 74, 77, 191
Persian Gulf, 89, 146, 148, 152, 157, 159, 162
PetroKazakhstan, 115, 158–159
Pharmaceuticals, 167–169
Pilgrim, Pilgrimage, 28, 207, 226, 237, 244
Primakov, Evgeni, 9, 17
Punjab National Bank, 167
Punj Lloyd, 157, 159
Putin, Vladimir, 9–12, 17, 31

Qatar, 145, 159, 227
Qing Empire, 215, 230

Raw material, 17, 106–110, 118, 166, 183
Refugees, 12, 36, 67, 88, 173–175, 207
Regional cooperation, 47–48, 62, 68, 70, 72, 76–77, 114, 174–175, 178, 185–188, 190
Regional Economic Cooperation Conference on Afghanistan (RECCA), 68, 70, 77, 186–188
Reliance Industries, 159
Roskosmos, 17
Rosneft, 149
Russia, Moscow, 1–4, 6, 9–22, 25–26, 28–36, 38–40, 63, 65–68, 70, 72, 75–77, 100, 104, 109, 115–116, 122, 126, 133–138, 142–143, 145–152, 155, 160–168, 183, 186, 190, 192, 210, 219, 236, 245

Sakhalin, 17, 151–152
Salafism, 217, 219, 227, 231
Salma (dam), 70, 180
Samarkand, 197, 199–200, 205–206, 210
Satpayev (block), 158–159
Saudi Arabia, 37, 68, 84, 227, 232, 238

Security, 3–4, 6, 10–12, 14, 16, 19, 21, 25–26, 28–30, 32–35, 37–38, 41–42, 45–47, 49, 52, 54–55, 62–65, 67–70, 72, 74–76, 82–84, 86, 88–89, 99, 100–101, 104, 121, 126, 129, 132–133, 135–136, 138, 149–150, 152, 164, 173–180, 184, 190–192, 224, 232

"September 11," 3, 12, 25, 28, 30, 32, 62, 68, 75, 82, 86, 88–89, 173, 240, 245

Serakhs, 162

Shaibanid (dynasty), 198–199, 205, 209

Shanghai Cooperation Organization (SCO), Shanghai Group, 4, 12, 15–16, 19, 25, 28–35, 37–40, 50, 52–56, 63, 67–68, 76–77, 89, 100, 121, 127, 136–137, 155, 184, 186

Shymkent, 158

Siberia, 16, 23, 34, 146

Sikkim, 98–99

Silk Road, 1, 5, 149, 198, 209, 215

Singh, Manmohan, 75, 113, 159

Sinicization, 28, 230

Sino-Pak (axis), 81–91

Sinopec, 33, 109–110, 157

South Asia, 15, 41–42, 44, 63, 70, 73, 84, 89–90, 122, 137, 161–162, 185–186, 189, 191

South Asian Association for Regional Cooperation (SAARC), 68, 70, 75, 77, 81, 89, 100, 104, 185–186

South Korea, 1, 3, 16, 23, 38, 101, 104, 125, 130

South Ossetia, 11, 14, 34, 143

South Yolotan, 33, 158

Soviet Union, 1–4, 10–11, 13, 16–17, 25–27, 29, 38, 43–44, 46–47, 61, 77, 85, 88, 100, 116, 121, 127, 147, 150, 228, 236, 246

Syria, 84, 108, 246

Tabligh-e-Jamaat, Jama'at al Tabligh, 227, 235, 237–247

Taiwan, 26, 28, 38, 99, 104, 105, 110, 125

Tajikistan, 3–4, 11–12, 21, 26–27, 29–30, 33, 35, 40, 48–49, 63–64, 67, 75, 81, 112, 123, 128, 133–134, 137, 155–156, 160–162, 164, 168–170, 186, 188–189, 194, 235, 237, 240–241, 245, 247

TALCO (Tajik aluminum plant), 164

Taliban, 4, 27, 29, 31, 49, 62, 64–70, 72–74, 76–77, 82, 85–89, 91, 173–174, 179, 180, 183, 191, 209, 232, 248

TAPO (Tashkent aviation factory), 20

Tarbagatai (Protocol of), 26

Terrorism, 3–4, 13–14, 16–17, 20, 25, 29–31, 34–37, 62–70, 75, 81–82, 85–86, 88–91, 126, 128, 130, 132, 136–137, 184, 232

Textiles, 166–167, 205

"Three evils" (doctrine), 29, 31, 36

Tian-Shan, 161

Tibet, 4, 14, 99, 102–103, 114, 129, 147–148

Timurid (dynasty), 199–201, 204, 206–207

Transneft, 149

Transoxiana, 200, 206–208

Transport, 49, 64, 70, 76, 134, 146, 156, 160–162, 166, 186–190

Turkey, 1, 3, 9, 13, 37, 68, 125, 141–143, 145–146, 148, 186, 188, 198, 219, 226, 235, 237, 240, 243–244, 246

Turkic, Turko-Mongol, 2, 27, 35–36, 40, 201, 215–216, 219, 229–231

Turkmenistan, 3, 11, 33, 35–37, 40–41, 63, 75, 81, 112, 134, 142, 145–146, 148–151, 155–156, 158–160, 162, 168–169, 186–189, 194, 235, 240

Turkmenistan-Afghanistan-Pakistan (pipeline), 20, 150–152, 159, 187

United Arab Emirates, 3, 9, 186

United Nations, 16, 111, 173–174, 177

United Nations Office of Drugs and Crime (UNODC), 177

Uranium, 5, 12, 17, 19–20, 89, 112, 125–126, 130, 151, 163–165

Urumqi, 32, 36–37, 90, 148–149, 152, 220, 226–227, 233

USA, Washington, 1–4, 9, 14–15, 18–22, 25, 28–29, 31, 36, 38, 43, 61, 63, 66–68, 72–77, 82–92, 98, 100–102, 104–105, 107, 110–111, 116–117, 120–122, 124, 130, 133, 135, 143, 147–148, 150, 156–157, 160, 173, 178, 182–183, 186, 192–193

Uyghur, 4, 6, 14, 27–29, 33–37, 39–40, 86, 128, 215–217, 221–233

Uyghur World Congress (UWC), 36

Uzbekistan, 11–12, 30–31, 33–34, 36–37, 63–64, 70, 75, 81, 123, 125, 128, 134, 148, 155–156, 158–163, 165, 168–169, 172, 186, 188–189, 194, 198–199, 218, 232, 235–237, 240, 243, 245, 247

Wakhan (corridor), 26, 88, 156, 161

War on terrorism, 82, 86

Wen, Jiabao, 30, 32

West Asia, 63, 70, 77, 89, 146, 185, 199, 205

World Bank, 110, 161, 186, 188

World Trade Organization (WTO), 17, 105–106, 110–113, 185

Xinjiang, 3, 5–6, 25–29, 31–39, 51, 66, 84, 86–87, 89–90, 124–125, 128, 147–149, 152, 155, 158, 160–161, 165, 170, 215–233

Yakub beg, 217, 233

Yang, Jiechi, 87

Zahir Shah, 61, 69, 174

Zhanazhol, 158

ZTE, 66